Howard Hospital and Infirmary for Incurables

Accidents, Emergencies, and Poisons

Keep this where you can Find it

Howard Hospital and Infirmary for Incurables

Accidents, Emergencies, and Poisons
Keep this where you can Find it

ISBN/EAN: 9783337172787

Printed in Europe, USA, Canada, Australia, Japan

Cover: Foto ©berggeist007 / pixelio.de

More available books at **www.hansebooks.com**

ACCIDENTS, EMERGENCIES,

AND

POISONS.

KEEP THIS WHERE YOU CAN READILY FIND IT.

DISTRIBUTED THROUGH THE

HOWARD HOSPITAL AND INFIRMARY FOR INCURABLES,

1518 AND 1520 LOMBARD STREET,

PHILADELPHIA.

———•◦•———

FOR SALE BY

JAMES HAMMOND,

No. 1224 CHESTNUT STREET.

1874.

INTRODUCTORY.

The object of this little book, as it is hoped every page will show, is to give a few directions as to what should be done in cases of the common accidents, emergencies, and poisons, until the arrival of skilled assistance.

The suggestions for the purpose are simple in character and everywhere readily found; and whenever useful, the nature of the difficulty as far as can be stated is explained in as few easily understood words as possible. A knowledge of this alone, without further information, will often enable a bystander of ordinary thoughtfulness to do all that is to be done in the shape of precautionary measures, to interrupt the unfavorable tendency of the trouble, if left alone.

It is written by a physician mainly for gratuitous distribution, and the entire expense of preparing the cuts and publication has been borne by a kind friend. Both of them will feel amply rewarded for the trouble taken in carrying out this intention, by knowing that what is here written has in any way tended to prevent unnecessary suffering, or has in a degree contributed to save the life and activity of some useful member of the community.

HOWARD HOSPITAL AND INFIRMARY FOR INCURABLES,

1518–20 LOMBARD STREET, PHILADELPHIA.

October, 1874.

Scarcely a month passes by, to many persons, without meeting somewhere, an accident or an emergency in which a little reliable information is not of the greatest service. One of the difficulties usually to be contended against in such cases is, loss on the part of the bystanders to know just what should be done. It will be found, as a rule, that the simplest things, usually the most useful, are neglected, while there is a disposition to rely upon cumbrous appliances, often of disadvantage, and sometimes, positively hurtful.

The object of the writer of this pamphlet is to present in a compressed form, for easy recollection and ready reference, a few suggestions as to what should be done in certain cases of emergency, until the arrival of skilled professional assistance. It is not saying too much, perhaps, that what is to be done to give relief or save life, in the greater number of cases, must be done by some one else *before* the aid of a physician can be procured. It has been truly said, "for want of timely care, millions have died of medicable wounds."

As far as possible the use of technical terms will be omitted, although where necessary they will be used with a brief definition enclosed in brackets; but the writer would respectfully suggest, that whenever possible the scientific terms should be remembered and used, instead of the popular expressions for the same thing. A scientific term the world over means but one thing, while a popular expression, in one place, means one thing, and in another, two or three things; and, possibly, nothing at all.

ACCIDENTS IN GENERAL.

An accident anywhere, if there are people about, assembles a crowd around the victim. The first thing to be done is to disperse it; or, at least, get the people to keep *away* from the injured person. A space of at *least* ten feet on every side should be kept *wholly free* from everybody except the *one or two* in charge of the operations for relief. If others are needed for a moment, to assist in some special duty, as lifting, removing of dress, etc., they can be specially selected from the crowd; and, having been of service, can immediately return where they came from. In several instances the writer has seen a person just removed from water, or gas, so closely surrounded by a dense mass of "relatives" and "friends," that it was impossible for the physician to freely use his arms. The *kindest* thing a bystander can do, is to *insist* upon a free space as large as suggested, and select from the crowd persons to hold themselves in readiness to start for whatever the physician or the individual in charge of the case may require. To show how little real interest the inside layer of the crowd usually takes in the restoration of the patient, it will often be found that it is almost impossible to get one of them to run an errand in the interest of the sufferer.

If the person has been thrown from a carriage, injured by a fall from a height, blow or other cause; while there may be no fracture, or other *external* injury evident, the nervous system has received what is called a "shock." As is commonly said, the person is "faint."

A person situated with such symptoms, should if possible, be placed flat on the back, with the head, neck, and shoulders *slightly* raised. The limbs, at the same time, should be straightened out, if practicable; so that the heart, already depressed in action, may act at as little disadvantage as possible. The cravat, collar, and everything else calculated to in any way impede the circulation toward the head, or the movements of the chest, should be loosened or removed. If the injury is slight, reaction will soon come on after giving the person a sip of cold water; brandy and water (teaspoonful in a tablespoonful of cold water every couple of minutes); or Aromatic Spirits of Ammonia (twenty drops in a tablespoonful of cold water) every couple of minutes. Gentle frictions to the extremities, a few drops of cologne water on a handkerchief to the nostrils; if the weather is hot, the use of a palm-leaf fan; hot flannels to the limbs and epigastrium (pit of the stomach); are all likewise useful in assisting reaction.

By this time should a surgeon have arrived, he will examine and decide upon the special nature of the injury, and inaugurate measures of special relief. Should he not have appeared, and it is thought best to remove the patient to the hospital, or his home, a *stretcher* should be secured, or a substitute, in the shape of a settee or shutter, provided. The injured person should then be gently *slipped* on, seeing that the body is supported as much as possible along its length, something thrown over or held over the face, to prevent, as much as possible, the uncomfortable feeling of being stared at in passing along. Four persons of uniform gait should then gently lift the stretcher and slowly carry the person to his destination. In the city of Philadelphia appliances for carrying injured persons are required to be kept at the station houses, and can be obtained, on application, as well as the services of a good policeman. The authority of the latter is almost invaluable in keeping away the crowd referred to, and in securing useful attention in conveying the person through the streets. If the person is to be taken to the Pennsylvania Hospital,*a dispatch from the Station House will secure, free of charge, an *ambulance* with competent persons to take charge of the injured individual.

Directions for fractures and dislocations are given elsewhere, p. 36.

ASPHYXIA.

This commonly-used word is from the Greek, signifying an *absence of pulse*. It states a fact, but not the cause. Like many other old words, the original meaning has been set aside, and it now means *suspended animation*, produced by the non-conversion of the venous blood in the lungs into arterial. The *supply* of good air to the lungs being cut off by some cause, the necessary *purification* at that point no longer takes place, and death of the entire body ensues from the absence of arterial blood, or the presence of venous blood; some physiologists regarding it from one cause, others from the other. In other words, as is often said, the person dies because the blood is not purified.

It will be seen there can be several varieties of asphyxia, as, (1.) Asphyxia from submersion, as in the ordinary drowning in water or other fluids; (2.) Asphyxia from mechanical causes, as by strangulation, or hanging, and by foreign bodies in the windpipe or its approaches; (3.) Asphyxia by inhalation of gases, known as suffocation; (4.) Asphyxia from torpor of the medulla oblongata (an important portion of the brain, at the junction of the spinal cord and what is called the brain), produced by the introduction into the blood of certain poisons.

Drowning.

As said above, this is *Asphyxia* from submersion in water or other fluids. This accident is of such frequent

* Or to the University Hospital, in West Philadelphia.

occurrence, and what is to be done must be done so quickly, that it is the duty of every member of the community to understand the measures of relief in such cases.

The body should be recovered as soon as possible from the water; the face turned downwards for a moment, with the forefinger of a bystander slightly curved and thrust backward, to depress the tongue, to favor the escape of a small quantity of water, or mucus, or other substances, often collected at the base of the tongue, over the entrance to the trachea (windpipe), which tend to obstruct the entrance of air to the lungs. The barbarous practice of rolling the person over a barrel, or hanging him head downwards, to permit the escape of water from the lungs, has almost ceased, in view of the fact, now generally known, that no water can get into the lungs.

This can be done as the body is being conveyed to the nearest house, a messenger having been previously despatched to make the arrangements involved in the following: — As soon as the body arrives it should be stripped of the clothing, rapidly dried, placed on a bed previously warmed, the head, neck, and shoulders, raised a very little, if any; frictions with the dry hands used to the extremities, and heated flannels kept applied to the rest of the body.

For an instant imagine the condition of affairs. Each atom of the body requires arterial blood, which is blood purified in the lungs by exposure to the air breathed. This *purification* has been *suspended*, and to that extent the *life* of the body is suspended. Movements of the chest, by which air is inhaled, are at a stand-still, and cannot, of themselves, be resumed. If *artificial breathing* can be

carried out for some time, it will be seen that these impurities may so far be removed that *natural* respiration can
take place.   Two methods are usually employed for the
purpose—the first known as "Silvester's Ready Method."

This consists, after the above suggestions have been carried out, in p u l l i n g t h e t o n g u e f o r w a r d, which
better favors the passage of air along the base of the tongue
into the trachea (windpipe), and t h e n d r a w i n g t h e
a r m s a w a y f r o m t h e s i d e s o f t h e b o d y
a n d u p w a r d, s o a s t o m e e t o v e r t h e h e a d,
by means of which the ribs are raised (expansion of

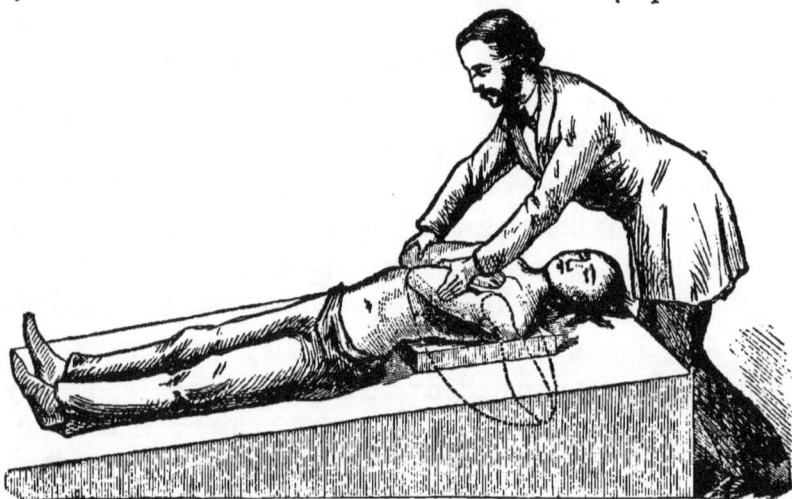

the chest) by the muscles (pectoral) running from them
to the arms near the shoulder.   A vacuum is thus created
in the lungs, the air rushes in, and the blood then is purified by the passage of the impure gases in the blood vessels to the air, and by the giving up by the air of a portion of its oxygen to the blood. T h e a r m s a r e n o w
b r o u g h t d o w n t o t h e s i d e s, a n d t h e
e l b o w s m a d e t o a l m o s t m e e t o v e r w h a t
is c a l l e d t h e "p i t o f t h e s t o m a c h." This
produces contraction of the walls of the chest, and expul-
sion of the impure air from the lungs.

These two movements constitute an act of respiration, and should be persisted in, without interruption, at the rate of about sixteen to the minute. In other words, each complete movement should occupy about four seconds, which is about the natural rate of respiration in health.

The second "Ready Method," as it is called, is that of Marshall Hall.

The person whose breathing is to be restored is placed flat on the face, gentle pressure is then made on the back, the pressure removed, the body turned on its side, or a little beyond that. The body is then turned again on the face, gentle pressure again used to the back, then turned on the side. This should be done about sixteen times in a minute.

Both of these methods have the same *object* in view; either may be exclusively used, or one may be alternated with the other. Most physicians express a preference for the first described ("Ready Method of Silvester)." Both of these procedures might be practiced, in advance, by the reader, because such practice might be more easily remembered than a concise rule. The application of the tourniquet, or pressure of the fingers, described elsewhere, to a blood-vessel, might also be practiced at the same time. There are few people, in an ordinary life, who will not find knowledge of this kind, at their fingers' end, of the greatest use.

In speaking of the restoration of persons drowned, it is often said that he was a good swimmer, and must have been attacked with "cramp." This is a spasmodic contraction of the muscles beyond the control of the individual, and occurs after exhaustion of the muscles from

over-exertion. Persons suffering from debility, especially
the debility peculiarly affecting the nervous system, should
never be induced to go beyond depth in the water, or out
of reach of immediate assistance. There is no warning in
advance of the seizure, and the person sinks at once. Many
lives are lost each season, in shallow as well as in deep
water, from these seizures, which could have been avoided
had the bather, perhaps just recovering from an attack of sickness, or even of indisposition, not neglected the
precautions stated.

Recovery from Asphyxia from drowning, can scarcely be
expected to take place after an immersion of five or six
minutes, although there are well authenticated cases where
restoration has taken place after an immersion of as much
as twenty minutes. The efforts ought to be made, and
persisted in until the arrival of a physician, or for at least
a couple of hours. As soon as returning *vitality* permits,
a few drops of brandy, in a little water, may be given;
and, as the strength of the person is usually completely
exhausted, from muscular efforts of the most violent and
continued character, to save himself from drowning, some
beef tea, or other easily-digested nourishment, should be
given.

Hanging.

Here death results from Asphyxia induced by pressure
applied to the trachea (windpipe) from the outside, as in
strangling, or hanging. The body, if hanging,
should be at once cut down, taking care
not to let it fall, so as to complicate
the trouble. If the knot can be untied readily, it
should be done; if not, cut the ligature. Remove, by
the finger, as in directions in drown-
ing, any accumulation of mucus at the

base of the tongue, place the body on the back, just as directed for a person taken from the water. If the body is still *warm*, after removal of the clothing, the face, head, neck, and chest should be dashed freely with cold water. To do this successfully, a person should stand off six feet, or even more, with a bowl of cold water, and then throw its contents, with as much force as possible, against the person. After using for a suitable length of time, the water should then be rapidly wiped off with a towel. There is little difference in essential features, after removal of the ligature from the neck, in the condition of a person who has been hanged or who has been drowned. In both it is Asphyxia: in one case, the air was kept from the lungs by a ligature, in the other, by liquids. Artificial respiration in both of them must be used, assisted for the same reason, and in the same manner, by like auxiliaries.

There is an impression, quite prevalent among the ignorant, that a penalty is incurred at law for cutting down the body of a person found hanging, unless the sanction of the coroner is obtained. No such delay is necessary, nor even justifiable; and an effort should at once be made to restore suspended animation by the methods given.

Suffocation.

There are several gases, which, when inhaled, are followed by symptoms of Asphyxia. The little valve (epiglottis) over the entrance of the trachea ("windpipe"), is so extremely sensitive that it will not even permit a drop of water to pass without a spasmodic closure of the opening, followed by coughing. It is not only sensitive to solids and liquids, but also to the presence of most gases. At one time it was thought that all gases were taken past it into the lungs, and absorbed from thence into the blood. The opinion now seems to prevail that most of them irritate the valve spoken of at the entrance of the trachea (windpipe), and closure of the entrance follows. The breathing is thus interrupted much as it is in drowning, where the liquid cuts off the passage of air to the lungs; or as in hanging, where the ligature prevents the entrance of air. In such cases death results from Asphyxia.

Carbonic Acid Gas.

Asphyxia, by inhalation of gas, takes place as soon as the person comes within the influence of this compound, and takes it in with the breath. A sudden sense of suffocation is felt, with dizziness of the brain, and inability to stand. If a person is standing at the time the air is taken into the lungs, and falls over, he is in a position, while down, to inhale more of the carbonic acid gas, for, being heavier than the air, much more of it is to be experienced at the bottom of the well, or cavern, than five feet higher up.

This gas, sometimes known under the name of "Choke Damp," is produced in the ordinary process of fermentation, in burning and slaking lime; and it is also found in mines, particularly coal mines, and in wells, cellars, or caves which have been long closed up. It is considerably heavier than the atmosphere, and is consequently found lying on the floor of the cavity where confined.

No well, vat, old cellar, or cavern of any kind, should ever be entered without first lowering down into the deepest point a lighted candle. If the flame is extinguished, or burns dimly, indicating the presence of this gas, no one, under any circumstances, should be permitted to enter without removing this foul air. It lies at the bottom, because too heavy to ascend. It is not so heavy, however, that a strong *current* of common air will not dislodge it. Buckets of water dashed down into the well, or masses of lighted shavings or blazing paper, give enough *movement* to the carbonic acid gas to dislodge it from its resting-place. After *testing* the success of the effort by again introducing the lighted candle, it can soon be known whether a person may enter with impunity. Freshly-slaked lime also rapidly absorbs it.

Often there may be no such gas shown in the *cavity*, but the efforts of the workmen will *dislodge* it from an adjacent space into the one in which he is breathing. This *possibility* should never be lost sight of.

When a person appears overcome by this Carbonic Acid gas, he is, of course, wholly unable to help himself, and he must at once be removed by another. Sometimes a grapnel hook can be used with advantage, but often the better way is to rapidly lower some bold, clear-headed person, with a rope securely fastened around his middle, who can seize and bring to the surface the unfortunate individual. No time should be lost in descending or arising, as the person lowered depends upon doing everything during the interval he can hold his breath; for, of course, should *he* inhale the gas, his position, in this respect, would be but little better than the man he attempts to succor. A large sack is sometimes thrown over the head and shoulders of the person who descends. It contains enough air to serve for several inhalations, while the texture of the material prevents, to a hurtful degree, the *admission* of the deleterious gas.

The person suffering from Asphyxia from the gas, immediately after being brought out, should be placed on his back, the neck and throat bared, and any other obstacles to the breathing quickly removed. His body should then be quickly stripped, and if he have not fallen into water on being overpowered by the gas, his head, neck, and shoulders freely dashed with cold water.

Remember, this is not "sprinkling," as commonly practiced; but, as said before, a person should stand off some distance, with a bowl of cold water, and *throw* its contents, with as much *force* as possible, against the parts. Others should follow without an interval for half a minute, while one can count thirty slowly, then the dripping water wiped away by a towel. This procedure should be repeated from time to time, as apparently required. Sometimes, if a brook of water is near, the stripped person might be dipped again and again; being careful, of course, not to dip in his face. Artificial respiration should be used with as little intermission as possible.

Should the person have fallen in the water and become *chilled*, the use of the cold water, in this manner, had better be avoided, as the evaporation of the moisture absorbs more heat than can be manufactured by the exhausted and overpowered system. In such a case, the body of the person should be put into a warmed bed, with hot applications, and Artificial Respiration (p. 10) at once established, as in the Asphyxia from drowning and hanging.

While artificial respiration is being used, friction applied to the limbs should be kept up.

Burning Charcoal

Certain gases (Carbonic Oxide Gas) are given off during the burning of charcoal, of a very poisonous character, and when inhaled for a sufficient length of time, rapidly prove fatal. The person quickly drops insensible, and dies of Asphyxia, in many respects like the person who has succumbed to the Carbonic Acid Gas, just described. The treatment there advised, should at once be carried out.

Anthracite and Bituminous Coal.

These also, when burned in a close room, as a kitchen shut up for the night with an open stove of these burning coals, give off, to a degree, the peculiar poisonous gas alluded to as coming from burning charcoal; Carbonic Oxide Gas; as well as *other* noxious gases. Persons sleeping in such a room, under the circumstances, unless awakened as the air becomes fouled, will be found senseless or dead, soon after. The treatment should be as described in the preceding pages, in Asphyxia from inhaling Carbonic Acid Gas.

Common Burning Gas.

Persons retiring at night often leave the gas "turned down," and the flame becomes extinguished. Enough gas often escapes to give trouble to the sleeper unless the room is well ventilated. Persons have been known to "blow it out," as they would a candle, and suffocation more or less complete has followed.

Treat as in the Asphyxia from other gases just described.

Foreign Bodies in the Throat.

A piece of food, or some other body, often gets back into the mouth, and cannot be swallowed. In such a case, the finger will often be able to thrust it downward, should that be thought best. A *hair-pin*, straightened and bent at the extremity, will often drag it out. If the body

2

is firm in character, a pair of scissors, separated at the rivet, and one blade held by the point, will furnish a *loop*, which often can be made to extract it.

Foul Air in Drains and Privies.

This is usually Sulphuretted Hydrogen, and arises from the decomposition of the residual matters usually found there.   Great caution, on this account, should always be observed on opening and entering such places, or places in possible *communication* with them, especially if they have been long closed.   A small quantity of *pure* Sulphuretted Hydrogen, if inhaled, is usually fatal; but, in the cases referred to, the gas usually exists *diluted* with common air.   The breathing becomes difficult, the person loses his strength, falls, becomes insensible and cold, lips and face blue, and the mouth covered with a bloody mucous secretion.

The person should be removed as quickly as possible beyond the influence of the foul air, and the treatment under the head of "Carbonic Acid Gas" pursued.

The *possibility* of such a disaster should always be borne in mind in opening long-closed drains or privy vaults, and the danger lessened by taking a few pounds of chloride of lime (bleaching salt), dissolving it in a pailful of water, and dashing it into the cavity.   In the absence of this, lime and water, in the form of the common "white wash," may be employed.   This gas readily combines with lime; to that extent freeing the air of the poisonous compound.

SUNSTROKE.

Ordinary exhaustion, from overwork in a heated atmosphere, is about the only disorder likely to be confounded with sunstroke.   In directions for popular use, like these, the distinction between the two will not be attempted, as there is no essential difference in the treatment.

Contrary to what is generally supposed, exposure of the head to the *direct* rays of the *sun* is not necessary, as statistics show it occurs in the shade, under shelter, and even at night; sometimes, even in persons who have not been exposed to the sun for days before. Intense *heat* always appears necessary to produce it; but the heat need not be *solar*, but may be artificial. Workmen in sugar refineries and laundries have been attacked.

Sunstroke appears to be decidedly favored by intemperance; want of acclimatization; and the debility which has been brought on by fatigue in a heated atmosphere, also favors it. Occupants of badly-ventilated sleeping apartments appear to be oftener attacked than those who sleep in purer air.

Symptoms.—It is generally thought by the non-professional, that the symptoms of sunstroke come on without any *warning* whatever. Most cases, however, are preceded by pain in the head; wandering of the thoughts, or an inability to think at all; disturbed vision; irritability of temper; sense of pain or weight at the pit of the stomach; inability to breathe with the usual ease and satisfaction. These symptoms become more marked until insensibility is reached, sometimes preceded by delirium.

The skin is very hot, usually dry, but when not dry, covered with profuse perspiration. The face is dusky, or, as the saying is, "blue;". breathing, rapid and short, or slow and sighing. The action of the heart, indicated to the hand placed over it, is weak, rapid and tremulous, often compared to the "fluttering of a bird." In many instances, from what is popularly termed the *commencement* of the attack until it ends in death, the patient does not move a limb, nor even an eyelid.

The breathing gradually fails; the blood, therefore, is not purified in the lungs, as is indicated by the livid, purplish appearance of the surface. We are led by it to conclude that death takes place by Asphyxia, as described under the heads, "Drowning," "Suffocation," etc., pp. 10–12.

Causes.—While we know certain things *favor* the disorder; that a *high* temperature is necessary to produce it; and advise certain measures of precaution and relief, found by experience useful in such cases; but little is known of the *nature* of the malady. It would seem that the grea᪴ heat of the body induces some change in the character of the blood, disqualifying it for the usual purposes of blood. From this peculiar condition of the blood, the portions of the brain or nervous system controlling the action of the muscles of the chest and heart lose their ability to superintend properly the movements of breathing and circulation, and, as said before, the person dies from Asphyxia.

Treatment.—T h e p e r s o n a t t a c k e d s h o u l d a t o n c e b e c a r r i e d t o a c o o l, a i r y s p o t, i n t h e s h a d o w o f a w a l l, or to a large room in a house with a bare floor; or, what is often better, if there is no sun, he should be placed in a *back yard*, on the pavement. Unnecessary bystanders must be kept at a *distance*, as the person in this, as in every other accident, needs all the pure air about.

T h e c l o t h i n g s h o u l d b e a t o n c e a n d g e n t l y r e m o v e d, a n d t h e p a t i e n t p l a c e d o n h i s b a c k, w i t h t h e h e a d r a i s e d a c o u p l e o f i n c h e s b y a f o l d e d g a r m e n t. T h e n t h e e n t i r e b o d y, p a r t i c u l a r l y t h e h e a d a n d c h e s t, d a s h e d w i t h c o l d w a t e r, i n p r o f u s i o n. While preparations are being made for this, a messenger should be despatched for a good supply of ice. A large fragment should be placed in a towel, and struck a few times against the side of the house, to rapidly reduce it to small pieces. These pieces, mixed by the hand into a bucket of water, will rapidly supply ice-water. Two buckets can be used, each half full of the small ice, and as soon as the water of one is used for dashing against the patient, another will be ready for the same purpose. The ice-water must not be *sprinkled* over the person, but *dashed* against him in large bowlfuls, particularly against the head and chest. While one person makes the ice-water, and another uses it, a third should, in the same manner, with a towel, break some ice in fragments not larger than almonds. A double handful, at least, of these bits should be placed in a thin, coarse towel, the ends gathered up and fastened with a string, as you would a pudding. Then holding to the tied portion of the collection of ice, the entire surface of the body should be rapidly *rubbed*. Indeed, two other persons might, each at the same time, be engaged at different portions of the body—not forgetting the head.

These measures are to reduce the heat of the body from the high temperature, evident even to the hand of a bystander, to something like a natural temperature. When the *decline* in the heat is noticed, the cold applications should be abandoned, the patient carefully removed to a dry spot, and the entire surface of the body dried off with towels. Should a tendency to a return of the high temperature be seen, as sometimes happens, even after consciousness is restored, it must be met by a renewal of the cold applications. The rise again in temperature need not seem surprising, when the amount of highly heated blood within the body, not yet exposed to the cold applications, is taken into consideration.

Artificial respiration, until the natural returns, must be resorted to as soon as the heated condition of the body is overcome. The dashing of cold water over the chest and face is a useful means of encouraging a return of the suspended breathing, and is practiced in asphyxia from other causes, pp. 10. The Ready Methods of p. 11, however, had better be relied on for this purpose.

Medicines in this malady, it will be seen, can be of little value. A stimulant, however, may be useful. Brandy, or any other form of alcohol, should be carefully avoided. The best stimulant in all such cases, if it can be obtained, is the Aromatic Spirit of Ammonia;* fifteen or twenty drops in a tablespoonful of water, might be given every few minutes, until taken three or four times.

Prevention.—During the heated term, as it is called, *all* use whatever of malt, fermented, or distilled drinks should be abstained from. Not only do they *favor*, in a general way, a condition of the system in many respects similar to that which leads to sunstroke, but they deaden sensibility at the very time it ought to be on the alert; and the person is less able to detect slight changes in his feelings, which otherwise might have served as useful warnings in his behalf. The use of such substances, under the circumstances, seems as unwise as it would be for a person, in a time of great danger, to prepare for watchfulness by taking a dose of laudanum; or for a worker with his hands among hot metal to apply something to them by which sensibility would be deadened or destroyed. By night, perhaps, he would have no fingers left.

Everything in any way calculated to impair the *strength* should be avoided. Sleep is a most wonderful restorer of strength, and the want of it is often caused by a badly-assorted late meal of the evening before. *Defective venti-*

---

* The Aromatic Spirit of Ammonia, and Brandy, quite independent of their intrinsic worth, are the two stimulants usually referred to, because, most likely of all others, to be found in an emergency. For the same reason, all through this treatise, but few simple appliances are directed, and these easily secured. It is an application of the principle of, one good thing, always to be had, rather than a dozen better, which cannot.

lation leads to a condition of affairs favorable to the malady under consideration. Every night a bath should be taken; but as this is not always possible in every house, the entire body should be washed off each night before lying down. Laboring men who work in the sun have no excuse for neglecting this, for water costs nothing, and three minutes of time is all that is required.

Drinking large quantities of *cold water*, merely because it is cold, should be avoided, particularly before, during, and after meals. The debility resulting from the heat weakens the digestive powers, and water unnecessarily used to excess at the times named tends still further to retard the digestion of the food, by further weakening the solvent action of the secretions of the stomach.

In other words, if there is a time above all others, the year around, when every precaution for the preservation of health is required, it is during the hot months of summer.

Loosely fitting light garments should be worn, if possible. Particular attention should be given the head. It should be *protected* from the heat of the sun, and at the same time the covering worn should favor the circulation of a free *current of air* over the scalp. A straw hat of loose texture, with a lining to the crown which could be kept constantly wet, ought to be worn ; and if it has brim enough to protect the neck, and even the shoulders, the wearer is just that much more fortunate than other people.

While attention should be paid to these things in hot weather, it is *particularly* necessary, should any of the named symptoms be observed on any special day, that the greatest care should be taken, if work in the sun is absolutely necessary, that the symptoms do not extend into an attack of sunstroke. Discontinuance of work, if possible, until the symptoms disappear, in such a case, would seem, to be the *best* course to be pursued.

It is said that persons who have once suffered from sun-stroke, for a long time after are unable to bear exposure to the heat, without a recurrence of the symptoms of the malady.

ACCIDENTS FROM LIGHTNING.

A person struck by lightning is usually rendered more or less *unconscious* by it, which lasts for a longer or shorter time.   Cases are on record where a person struck exhibited no sign of life for an hour, and then recovered· Temporary paralysis of a portion of the body may remain for a while, as well as disturbance of some special function, as the sight, smell, or taste.

The *burns* caused by lightning should receive the same attention as a burn from any other cause.   Sometimes an injury observed is not directly due to the electricity, but from a fragment detached by that agent from a neighboring substance.

When death takes place, it is from shock, as it is called, to the brain and nervous system.   When the person exhibits little or no signs of life, the clothing should be rapidly and immediately removed, the body exposed to a dashing of cold water; then dried, placed in bed, and warmth applied, particularly to the "pit of the stomach," by means of bottles filled with hot water, or the tin vessel kept in some households for such application.   It is somewhat concave on one surface, filled with hot water, and, if it can be had, is well adapted to the purpose.

Artificial Respiration should be kept up until the parts of the brain and nervous system in charge of this duty shall have recovered enough to attend to it.   As said before, recoveries after an hour of supposed death are on record.

Some stimulant, as the Aromatic Spirit of Ammonia,

may be used.   Twenty drops, in a tablespoonful of water, every few minutes, may be given ; or a teaspoonful of Brandy instead.

SHOCK.

Mild forms of Shock, or Collapse, as it is sometimes called, are often, by the non-professional, confounded with Fainting (Syncope).   As far as the symptoms extend, the symptoms of an ordinary attack of Fainting are analogous to those of Shock.   The symptoms between the two vary rather in degree and duration than in kind.

Life may be destroyed by certain agencies, as a blow upon the "pit of the stomach," or a sudden and powerful emotion of the mind, and no visible trace be left in any part of the body.   It is called "Death from Shock."   This is the *extreme* result of Shock.   Usually the patient lies in a state of utter prostration.   There is pallor of the whole surface; the lips are bloodless and pale.   The eyes have lost their luster, and the eyeball is usually partially covered by the drooping upper lid.   The nostrils are usually dilated.   The skin is covered with a cold clammy moisture, often gathered in beads of sweat upon the forehead.   The temperature is cold, and perhaps the person shivers.   The weakness of the muscles is most marked ; as the phrase is, "the patient is *prostrated.*"   The mind is bewildered, often insensibility occurs, unless aroused; and, in many cases, nausea and vomiting.   In *extreme* cases the nausea and vomiting are not so apt to occur.

Sudden and severe injuries, particularly if *extensive* in character, and involving a large amount of texture, cause Shock.   Burns—especially of children—extending over a large *extent of surface*, even if not extending to a great *depth*, are often followed by Shock, and this complication requires often the *earliest* attention.

Certain poisons, as Tobacco and Tartar Emetic, act in this manner, depressing the system. So does a current of electricity, as is seen in the effects of lightning.

Loss of Blood produces or aggravates Shock. Hence a *slight* injury, with much *loss of blood*, may be attended with more Shock, than a comparatively more severe injury *without* the loss of blood. Debility favors the influence of Shock. A weak system is more easily impressed by it, and, as should be expected, *reaction* from its effects is longer in taking place.

As the vital powers of life decline, from engrafted or natural causes, there is less power available as a reserve to meet contingencies. In youth there is an available fund of this kind; in the adult the resources of the system may be equal to the task of *ordinary* maintenance, but in the aged, as said before, there is much less ability to deal with *sudden* losses of strength. The *aged* are, therefore, slow to *rally* from the effects of Shock. They have more power of *resistance* than the young. The shock does not *readily* make an impression, as it does in the *young*, but when it *does*, the impression *endures*. In the young the impression is more easily *made*, but sooner *subsides*.

Treatment of Shock

Consists in first placing the patient as flat on his back as possible, with the head *raised not over an inch*. This is an important point in cases of ordinary Fainting, and whenever the vital powers are depressed. Stimulants are required. The aromatic character of Brandy enables it to be retained by the stomach when whiskey and other forms of Alcohol are rejected. A teaspoonful in a tablespoonful of water every minute, until six or eight have been taken, is the best way to give it. If the temperature of the body is *raised* by it, and there seems a revival of the action of the heart, *enough* Brandy has been given. Twenty drops of the Aromatic Spirit of Ammonia in a teaspoonful of water may be given every couple of minutes, until four or five doses have been taken. The applications of heat to the extremities and "pit of the stomach" are very useful. Flannels wrung out in hot water, or bottles of hot water properly wrapped up, should not be neglected. In some households, a tin can, somewhat concave on one surface, to fit the curvature at that point, and with a stopple in the upper surface for the introduction of the hot liquid, can be usefully employed for heat to the epigastrium (" pit of the stomach "). Mustard plasters to the same place are often used, but they are so inferior to heat for the purpose, if that can be applied, and so apt to blister, thereby making it impossible to use anything else on the surface, that some reluctance is felt in advising them. Nausea and vomiting often are seen in Shock, and can best be allayed by getting the patient to swallow whole, small chips of ice. Ice, by the way, can be easily chipped by standing the piece with the grain upright, and splitting off a thin edge with the point of a pin.

Ammonia (smelling salts), applied to the nostrils is often useful; and cologne, on a handkerchief, is often pungent enough to be of service in the same way.

FAINTING.

Persons often faint without any proportionate cause. Debility of the nervous system *favors* it. While the author would not like to say that the tendency to swoon can be intentionally acquired, he is compelled to think it can be unintentionally perpetuated under many circumstances.

The treatment *usually* followed is, perhaps, the best; but people are apt to *raise* the *head* of the patient. Even in carrying her to the bed or sofa, it should be kept lower than the rest of the body. Indeed, there is no better restorative in such cases than such a relative position of the extremities. Should the person be sitting in a chair at the moment, do not remove her, but stand behind the chair, reach the hands over in front, so as to grasp the sides of the back of the chair, take a step backwards to give room, and then slowly depress the back, supporting the head until the floor is reached. An assistant, by holding to the dress over the knees, will prevent lateral slipping off from the seat of the chair. It is so rapidly and easily done, besides so effective in its operation, that little else remains to be done. Usually the back of the head of the patient scarcely reaches the floor before consciousness returns.

Shock from Bathing in, or Drinking Cold Water.

In the hot weather, cases often occur where death or great prostration ensues from drinking ice water, or bathing in cold water, while the body is exhausted from heat or exercise. The same thing happens to animals under similar circumstances. Cold water in hot weather, if the person is *heated*, should always be drunk in small quantities at a time. If not, although neither death nor prostration may follow, a troublesome derangement of the digestive tract ensues, often laying the foundation for other troubles.

When the body is heated, or exhausted, a bath in cold water ought never to be taken. A sponge bath will answer, until the vigor of the body has had time to be restored.

These troubles can be referred to Shock, and should be *promptly* treated as such, according to the given directions.

BURNS AND SCALDS.

These common accidents, by receiving *early* and *suitable* attention, are often deprived of much of their inconvenience. Of course, the first thing is to put the fire out, and then, if the injured parts require it, the clothing should be cut away, so as to get at the entire extent of the injury with as little trouble to the patient as possible. Should any fragment of garment appear adherent to the burned surface, the *sticking* part should be left, as the violence required to remove it must necessarily *increase* the damage to the injured part.

When the clothing catches fire, throw the person down on the ground, as the flames will tend less to rise towards the mouth and nostrils.   Then, without a moment's delay, roll the person in the carpet or hearth-rug, so as to stifle the flames, leaving only the head out for breathing. If no carpet or rug can be had, then take off your coat and use it instead. *Keep the flame as much as possible from the face, so as to prevent the entrance of the hot air into the lungs.* This can be done by beginning at the neck and shoulders with the wrapping.

If the burn or scald involves considerable *surface*, symptoms of shock are observed, from the extreme of mere weakness to that of utter prostration. This at *once* requires prompt attention, and a few drops of Aromatic Spirit of Ammonia in water, or a little Brandy, should be given every few moments until a *return* of the strength is seen. A burn superficial, as far as *depth* is concerned, but covering a large *surface*, especially in the case of small children and aged people, is usually considered more dangerous, as far as *life* is concerned, than a burn *smaller* in extent, but deeper and more complete. Never mind how slight the injury *appears*, if there is reason to suppose the heated air or steam has been *inhaled*, no time should be lost in taking the opinion of a physician as to the result of the injury to the *throat* and *lungs*.

If the burn or scald is *slight* in character, one of the best applications is the Water Dressing, p. 50, as there said, keeping the linens used, *constantly* wet with cold water. In a short time after the pain shall have moderated, one of the best things for use, and readily procured, is a dressing of pure hog's lard. The common lard of the stores will scarcely answer, from the impurities it contains. At the Infirmary, in Lombard street, they usually direct that a half pound, or less, of the best lard should be bought, and put into a vessel of hot water, boiled a few moments, being stirred with a stick until it is thought all the salt used for preserving the lard, and the alum put in to bleach it, have been washed out. The vessel is then set aside until the floating lard hardens. It is then collected, placed in a bowl, which in turn is placed in a vessel of hot water on the stove, and kept there until the water mechanically held by the lard has been driven off. The bowl of lard is heated by surrounding water, to prevent *scorching*.

After thus prepared, the lard may be considered perfectly pure, and can be put away in suitable vessels until required for application.

It is much better than the commonly used Linseed Oil and Lime Water, as the Linseed Oil is rarely *pure*, but contains irritating substances left in the manufacture, or added for the purposes of adulteration. Wheaten Flour is often dusted over the burn; but this, with the discharges, *hardens*, and is of as little comfort as an application of small crusts of bread would be to the injured part.

Cotton wool (carded cotton, cotton batting) is often used, but the fibers become imbedded in the discharges, and then cannot be detached without unnecessary disturbance of the wound.

If the burn or scald, particularly the latter, is superficial in character, a simple and useful dressing is the application, by a brush, or a soft wisp of old muslin, of the White of Egg to the injury. As soon as the first layer dries, another should be used.

A lather of soap from the shaving cup, applied by the brush in the same way, is often followed by immediate relief. These substances appear to protect from the action of the air, the irritated nerves beneath. As before said, do not apply "cotton" to the injury, as sooner or later it increases the pain, and without having done any special good.

If a physician has been sent for, it is better not to make any domestic applications to the burned parts. Such things frequently prevent him from using those better adapted to the wants of the person, and keep him, too, from forming a correct estimate of the injuries.

Where the effects of the burn or scald extend *deeper*, involving the subcutaneous tissue, or even the parts beneath that, as the muscle, other considerations must not be overlooked. There probably will be more shock. The portion whose vitality has been *destroyed* by the burn cannot do otherwise than become detached from the uninjured parts beneath, thrown off in the shape of shreds or larger masses, during the process of sloughing. After water dressing has given a degree of relief to the part, and this is sometimes sooner secured by adding Laudanum (Tincture of Opium) to the water, a system of Poulticing* must

* "Recipes for Sick People," published for gratuitous distribution by the Howard Hospital and Infirmary for Incurables, 1518-20 Lombard Street, Philadelphia. Sometimes bound with this.

be commenced. After being used for a short time, a mark of well-defined *separation* is seen at the junction of the burned and the unburned parts. The edge of this dead portion often falls away, like a piece of wet buckskin, showing, except at the *edges*, a union of the dead and the living parts. This process of separation continues for some time, attended with such profuse discharges that the poultices must be *changed* several times a day to preserve neatness; but after a while the *entire* mass lies loose, attached only at the center to a raw suppurating surface below. A short time after the *whole* mass becomes detached and is removed. Then at the edges and surface of the uninjured parts a process of filling up the wound by "granulation" must commence. The newly-formed substance begins first to be deposited at the edges of the wound, generally reducing the area. This process, in such a wound, the result of a burn, is much less rapid than might be supposed by the unprofessional, and is attended with much suppuration (making of pus).

Whenever the destructive process of *suppuration* goes on in the body, from whatever cause, there is *exhaustion* of the strength. This must be combated by a judiciously selected diet; and sometimes by appropriately selected tonics. Often, the surface undergoing repair is benefited by local applications; but these can only be appropriately selected by a physician, so nothing need be said about them here.

As remarked above, in burns beyond a certain degree of destruction, the process is one of repair rather than restoration. Instead of the *destroyed* portion being *replaced*, the reparative material is of *such* a character that it undergoes contraction; and great deformity may result from its dragging effects upon adjacent healthy parts. These effects may often be *mitigated* in extent, but cannot be wholly *prevented*. Thus, if the arm, at the elbow, is burned or scalded, so that a scar results, the contraction of this tissue will often draw up the forearm to a right angle, from which it cannot be straightened. A burn or scald at the front of the neck is often followed by a dense white scar, which, contracting, draws the chin down towards the chest, and the lower lip down towards the chin, ending in the greatest deformity. The medical attendant is sometimes unjustly censured for these things.

From what has been said, it must be observed that Burns and Scalds practically differ but little from each other. Scalds are usually more confined to the *outer* cuticle, unless the substance containing the heat is viscid in character, as oil, pitch, etc., and does not rapidly run off the part with which it came in contact. As far as popular assistance is concerned, the two may be regarded as presenting no essential difference.

Burns by Lime, Caustic Potash and other Alkalies.

As a rule, these are troublesome, since there is not only removal of the cuticle, superficial skin, but destruction of the soft parts *below*. Lime is a powerful *alkali*, and rapidly destroys the parts with which it comes in contact. It is useless to attempt to *pick* it off, for the fingers remove no more than it gets hold of, so an application should at once be made of something to *unite* with the Alkali, to form a comparatively *harmless* preparation, v i n e g a r d i l u t e d w i t h w a t e r; the a c i d in L e m o n- j u i c e, o r a n y o t h e r d i l u t e A c i d, will answer. These things do not *undo* what has been *done*, they only prevent *further* mischief. The portion of the tissue *already* destroyed, must *separate* as if it had been destroyed by *heat* in the case of a Burn or Scald; must be aided by the same means; must heal in the same manner, and must be

3

. followed, of course, by the same ultimate contraction of the reparative material. And what has been said about the Alkali known as Lime, may be said about the *other* Alkalies; P o t a s h , S o d a , A m m o n i a , etc.

Burns by Acids—Sulphuric Acid (Oil of Vitriol), Nitric Acid (Aqua Fortis), etc.

As *Alkalies* destroy the living tissues they come in contact with; so will *Acids* of sufficient concentration. In such cases, a p p l i c a t i o n s o f w a t e r w i l l d i l u t e t h e m beyond their capacity to injure. Alkalies applied neutralize acids into *harmless* preparations. C o m m o n E a r t h , gathered almost anywhere, a p p l i e d i n h a n d f u l l s , contains alkali enough of one kind or another to entitle it to the consideration of being o n e o f t h e b e s t (and at the same time most easily secured) applications in cases of Burns by Acids.

CONTUSIONS.

These common injuries are termed "Bruises" by some people, and are the only other injuries beside wounds and fractures, produced by blows or pressure. The injury may be the *simple* form; only a slight shaking or jarring of the texture, with no *visible* change, except what results from the rupture of the blood-vessels. This is the most frequent. In the more *severe*, but less frequent form, the Contusion means broken blood vessels, muscles, and tissues between and around them; the parts are thoroughly crushed, sometimes to a pulp; damaged beyond recovery, and ready to perish in the gangrene resulting as the extreme form of such an injury.

The *quantity* of blood escaping from the ruptured vessels depends, in a large degree, upon the size and number of the vessels injured, but in a larger degree upon the space into which the blood can accumulate. A single divided vessel in the *scalp*, owing to *looseness* of the tissue there in which the vessels are distributed, may permit a swelling, the result of the escape of blood, extending in area over a half of one side of the head.

In Contusions, the first conspicuous symptom is that of *Shock*, which generally, but not always, bears a relation to the extent of the injury.   Thus a crushed finger is attended, as a rule, with much less shock than a crushed hand or foot.   Contusion of certain parts, as the larger joints, breasts, and other portions of the body, are followed by most *severe* symptoms of shock.   The *pain* is not always as severe as might at first be thought, for it is said the nerves are so much injured as to be deprived of their ability to receive and transmit the necessary impression. The *swelling* depends, at first, largely upon the *blood* poured out by the injured vessels, and as just said, this depends upon the *number* and *size* of the divided vessels, as well as upon the *character* of the part containing them.

Treatment.

In the milder Contusions, there is but *little* shock. Should there be more, p l a c e t h e p a t i e n t o n t h e b a c k, h e a d n o t e l e v a t e d, a n d g i v e s t i m u - l a n t s as directed. See Shock, p 25. The next thing is to limit the consequences *likely* to ensue from the ruptured blood-vessels.   This is best done by lessening the *supply* of blood to the part by *elevating* it, if possible, above the heart, and using c o l d a p p l i c a t i o n s i n t h e s h a p e o f p o w d e r e d i c e, tied up in towels, to the part; and along the course of the larger vessels going to the injury.

A large piece of ice secured in a towel, so the fragments cannot escape, can be reduced to fine fragments by a blow or two against the wall.   After it has remained on for a time,

the water may be substituted in the shape of a drip; * or
several thicknesses of wet towel may be applied, only they
must be dipped in cold water, squeezed out, and changed
every sixty seconds.   If not changed, the wet towels really
act as *poultices* to the part, *inviting* what we should try
to *prevent*.   When the Surgeon appears, special measures
will be directed by him.   Recollect it takes a great deal of
heat to convert ice into water, and water into vapor, and
if the patient has not got this heat, s y m p t o m s o f
c h i l l i n e s s w i l l b e o b s e r v e d.   When this happens the application must be *stopped* and the moisture
must be taken up by a towel; particular attention always
being paid to keep the bed clothing, and everything else,
perfectly dry and neat.

Discoloration is due to the color of this escaped blood,
seen through the cuticle, and varies from blackness, usually
indicating intense injury, so that the blood itself is poured
out, through dark blue, purple, crimson, down to delicate
pink, indicating only a blood-stained fluid.

After *preventing* the escape of blood from the vessels,
as far as *practicable*, there remains to get *rid* of what has
already been poured out.   In some forms, the assistance
of professional advice will have been secured.   In simpler
cases, the blood, at this later stage, should be encouraged
to flow through the now repaired vessels and the neighboring vessels as much as possible.   By this means the diseased particles are, as it were, picked up, as physicians say,
absorbed, and replaced by healthier ones.   The color of the
part gradually fades in proper time, and the injury is said
to be restored.   Gentle frictions of the part and neighbor-

---

* A pitcher, or some other vessel of water, placed higher than the limb,
with a moistened string or strip of linen. The end of the string is placed
in the water, the other hangs down on the outside, so the water will drip
along the string from the vessel to the injured part.

hood, the application of dry heat, and stimulating lotions, as Alcohol, Camphorated Soap Liniment, with other.simple things of the same character, are often used with the intention of *assisting* this process of nature.

FRACTURES AND DISLOCATIONS.

It is often evident to a bystander that a Fracture or Dislocation exists, without knowing what can be done in the interval which must elapse before the arrival of competent professional assistance. Of course no one but a very ignorant and bold man would attempt to do more than make the sufferer *comfortable* in the meanwhile.

In instances of suspected Fracture or Dislocation of the lower extremity, the injured parts s h o u l d b e p l a c e d in a c o m f o r t a b l e p o s i t i o n, a n d a s w e l l s u p p o r t e d a s p o s s i b l e, to prevent the *twitchings* of the leg from the spasmodic action of the muscles of the injured extremity. If necessary to remove the patient to his home or the hospital, from the spot where the accident happened, the arrangement of the limb should be made after he has been placed on the stretcher or substitute.

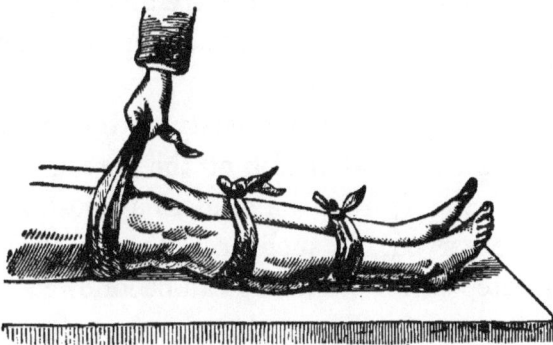

If found necessary to carry the injured person some distance, and a litter for the purpose cannot be had, the arrangement of the fractured limb against the other, and kept there by handkerchiefs, as shown in the cut, is often of great comfort to the sufferer.

If the general character of the injury is evident, in sending for the surgeon it is best to tell the messenger, so that, as far as possible, the necessary appliances can be provided before leaving the office. .

In the meanwhile, under no circumstances, should the bystanders be permitted to handle the affected part beyond what is absolutely necessary. As a general rule, a much longer time than is commonly supposed, by most people, may pass between the occurrence of the accident and the arrival of the surgeon without serious injury to the patient or ultimate disadvantage to the fracture. Many persons, thinking that the broken bone must immediately be "set," are apt to accept the services of the first person arriving asserting himself qualified to do it. Such an individual necessarily makes a more painful examination than is necessary, applies the splint—perhaps not at all the most useful—which the surgeon, arriving later, is obliged, out of consideration for the condition of the sufferer, to acquiesce in.

If the injury is to the upper extremity, the part should be placed in a supporting sling, and kept in a comfortable position.

Sometimes, owing to the severity of the injury, or the condition of the general health of the person at the time, symptoms of Shock, from the mildest expression to insensibility, are observed. In such a case the measures of treatment suggested under that head can be followed until other advice is obtained (p. 25).

WOUNDS.

For systematic study, wounds may be classified according to their direction, or depth, or locality; but for our purpose they may be arranged after the mode of their infliction : (1.) Incised wounds, as cuts or incisions, including the wounds where portions of the body are clearly cut off; (2.) Punctured wounds, as stabs, pricks, or punctures; (3.) Contused wounds, which are those combined with bruising or crushing of the divided portions; (4.) Lacerated wounds, where the separation of tissue is effected or combined with tearing of them; (5.) Poisoned wounds, including all wounds into which any poison, venom, or virus is inserted.

Any of these wounds may be attended with excessive *hemorrhage* or *pain* or the presence of dead or *foreign* matter.   As all wounds tend to present several *common* features, a few words will be said about them before describing the distinctive characteristics of each.

The first is h e m o r r h a g e (bleeding).   This depends, as to *quantity*, upon several conditions, the chief of which is the *size* of the *blood vessels* divided; and, to a degree, upon the *manner* in which it has been done.   A vessel divided with a *sharp* instrument presents a more favorable outlet for the escape of blood than one that has been divided with a *blunt* or serrated instrument, or one that has been *torn* across.   Except in the first named, the minute fringes or roughness necessarily left around the edges of the vessel at the point of division *retard* the escape of blood, and furnish points upon which *deposits* of blood, in the shape of clots, can take place.   Hence, all other things being equal, an Incised wound is usually attended with more *hemorrhage* than Contused or Lacerated wounds.

Personal peculiarities of the patient, and the health or disease of the wounded part of the body, may exert much influence upon the hemorrhage.   Usually it ceases in a short time by the coagulation (clotting) of the blood in the severed extremity of the vessel, without further attention than the application of cold, which favors *con-traction* of the blood ves-sel divided, as well as those leading *to* the injured part. Should an *artery* or branch have been divided (indicated by a *spurting* of a spray of bright blood at each beat of the heart), the bleeding may not cease at once. To stop it, the firm pressure of the finger for some time to the point of division should be used, to diminish the size of the vessel at that point, until a clot is formed there.

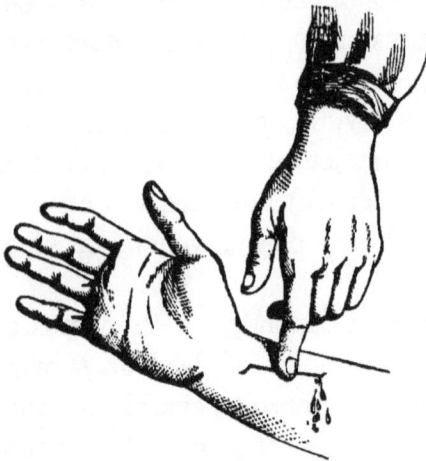

Sometimes, pressure to the supposed seat of the injured vessel does not *reach* the artery. In such a case the pres-sure must be used to some known trunk between the ori-ginal supply of the blood and the injured branch. Thus, if the finger or the toe is the seat of the arterial hemor-rhage, firm pressure applied each *side* of the finger, close to the hand (as in the cut) or toe, close to the foot, compresses the arteries pass-ing along to be distributed to the extremity. If the hand or foot is the seat of injury, pressure on the wrist, over the point where the ar-tery is felt for the "pulse," or at the inside of the ankle, will materially retard the passage of the blood beyond

those points. Should pressure by the thumb at these suggested points not answer the purpose, the main trunk of the artery, higher up, should be compressed by a tourniquet. Before this is done, it is always well to place the person injured flat on his back, and hold the arm and hand in a perpendicular position for a time, as the heart will then be unable to throw the blood with its usual *force* to the extremity. Pressure applied by the fingers, with broken ice in a towel bound round the arm, in conjunction with the elevation of it, will often stop the hemorrhage, or retard it, until professional aid is secured. If the foot is the seat of the injury, elevate the whole limb in the same way, applying pressure and pounded ice on the same principle.

In wounds of the scalp there is usually much loss of blood, owing to the abundant blood supply of that part. The firm skull below offers a good point for pressure, and the vessel rarely fails to be compressed if the thumb is applied over the point of division of the severed vessel.

The *amount* of blood actually lost is apt to be much over-estimated. Quite a *small* quantity will seem " a half pint" if distributed over the clothing, and a gallon of water requires no great amount added to it to give it quite a blood red color. It is estimated that about one-eighth of the weight of the entire human body is blood; in other words, the quantity of blood in a human body weighing 144 pounds would be about 16 or 18 pounds. Of course, this amount, nor half of it, perhaps, can be withdrawn from the vessels without fatal results; but it is merely mentioned to show that the entire quantity asserted to exist by physiologists is *much larger* than is popularly supposed. When hemorrhage from a divided blood vessel is seen there is usually much more apprehension and excitement about it than is warranted.

This figure shows the method of exerting pressure by the fingers along the course of the Brachial Artery; between the divided vessel and the heart.

If the wound should be in the arm, above the point indicated by the fingers, or in the axilla ("arm-pit"), pressure could be made by the thumb, a blunt stick, properly protected, or the handle

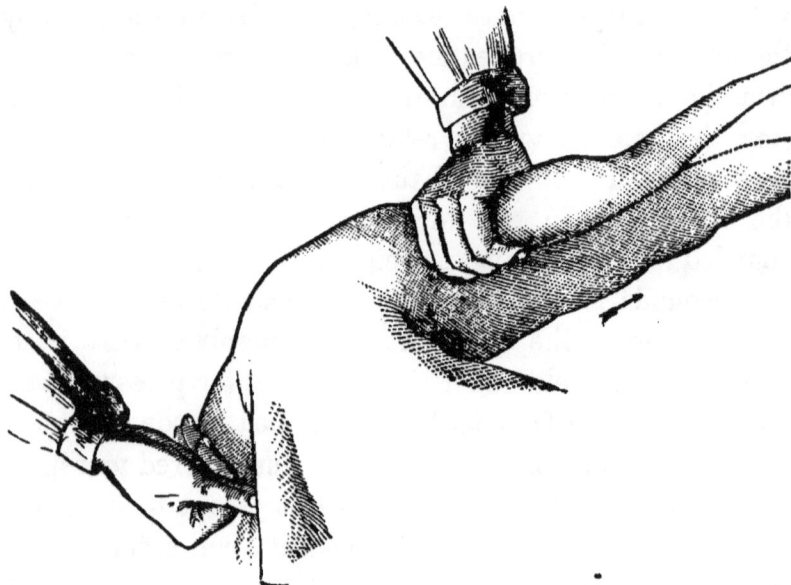

of a door key upon the Sub-clavian artery, which passes, as the name suggests, along under the clavicle ("collar bone") and down the arm, where it is called the Brachial Artery—just spoken of. Further down the arm at the elbow, this vessel is sub-divided into two others, each following a bone of the forearm to the wrist. At the wrist, over one bone, near the surface, the pulsation of the heart is sought by the finger of the physician.

NOTE.—The arm and forearm, with dotted lines, indicate the course of the arteries, and points at which pressure can be most judiciously applied.

The arrow points the course of the current of the blood of the artery, from the heart to the extremities.

Permanent pressure being exerted by means of a temporary tourniquet to the Brachial Artery spoken of on the other page. A common folded handkerchief, with a f i r m , s h a r p l y - d e f i n e d k n o t tied at the middle, a long strip of muslin torn from a shirt sleeve, or a suspender, with a suitable knot in it, is rather loosely tied around the arm, and the slack taken up by

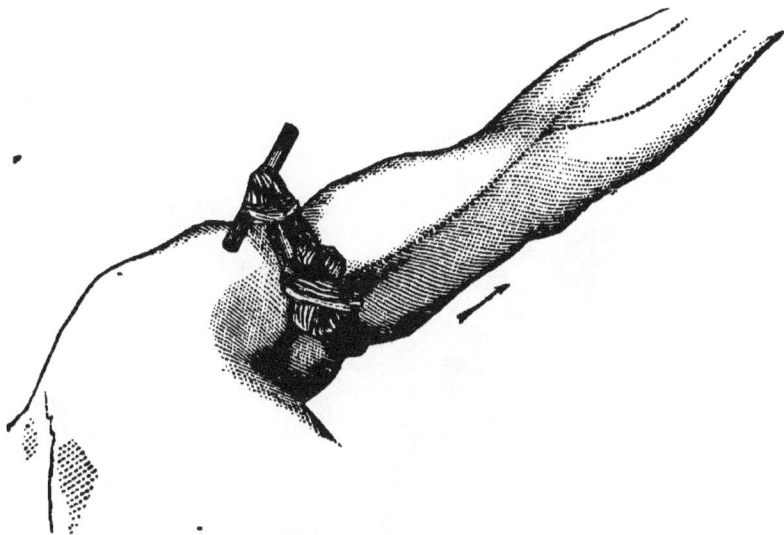

twisting with a cane or stick until the knot, kept over the vessel, exerts enough p r e s s u r e to prevent the passage along it of the blood.

This is easily done if you proceed to it quietly, without talking; especially if previously practiced once upon the extremity of a friend.

NOTE.—The arm and forearm, with dotted lines, indicate the course of the arteries, and points at which pressure can be most judiciously applied.

The arrow points the course of the current of the blood of the artery, from the heart to the extremities.

The method of exerting pressure by the fingers along the course of the Femoral Artery, between the wound and the heart.

Sometimes it is easier to find the artery nearer the surface, at a point along the dotted line, or a little higher up towards the groin. The two thumbs placed together furnish firm resistance; and a blunt stick, suitably protected, will often answer to keep up the pressure until a tourniquet can be extemporized.

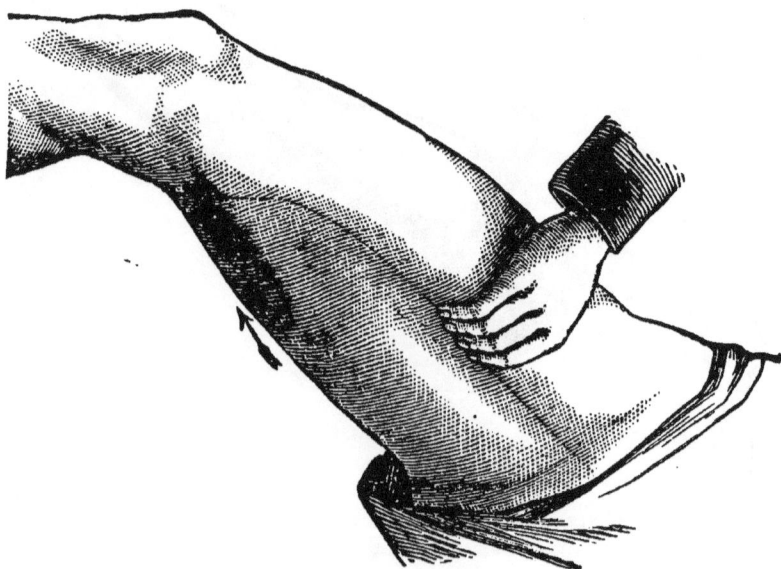

The muscular condition of the entire leg does not permit the pressure of the fingers to be as successfully exerted along the main arteries, as in the case of the arm just spoken of.

NOTE.—The thigh and groin, with dotted lines, suggest the course of the large arteries, and point at which pressure can be most successfully used.

The arrow indicates the direction of the current of the blood of the artery, from the heart to the extremities.

This cut presents the tourniquet made as directed on page 43, by getting a large firm knot in a handkerchief, or anything else of the kind. A small pebble has often been introduced for the purpose, into the knot, with success. Twist the ligature with the leverage obtained by passing under it a cane or stick.

Get the knot over the artery—keep the knot there, and tighten until the pressure of the knot closes the vessel.

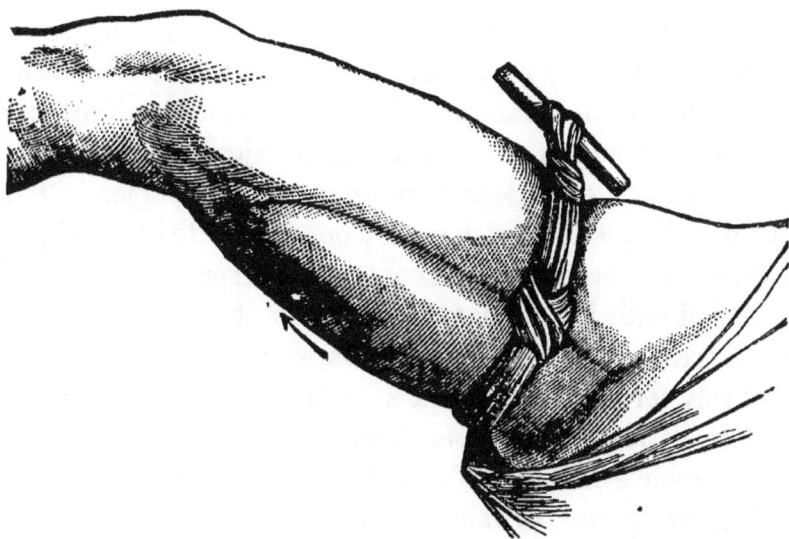

It is much easier done than imagined, especially if the individual has some day spent three minutes practicing the preparation of the ligature, and its application over the course of the artery.

NOTE.—The thigh and groin, with dotted lines, suggest the course of the large artery, and point at which pressure can be most successfully used.

The arrow indicates the direction of the current of the blood of the artery from the heart to the extremities.

There is no necessity for the alarm often shown, especially as it obscures the judgment of those who, if they would but reflect a moment, could much more serve the true interests of the sufferer by keeping cool and collected.

P a i n, it may be said, accompanies all wounds, for it is almost impossible to sever a blood vessel without severing nerves. It is usually much less severe than might be thought, and as little can be done immediately to relieve it, other prominent features of wounds in general will be spoken of.

F a i n t i n g, after severe hemorrhage, or in "nervous" persons, frequently requires attention, after the loss of blood has been placed under control. Often it is due to the *sight* of the blood, and an undefined apprehension as to the extent of the injury on the part of the wounded person. The latter feeling is in part derived from the excited and frightened appearance of those about. A person with a wound attended with hemorrhage, ignorant of its extent and consequences, seeing his friends, upon whom he must necessarily rely for succor, in such a state of alarm that he can expect little real aid from them, cannot be said to be in a comfortable state of mind—and is apt to faint.

The symptoms of fainting are too well known to need description here, especially as something is said about them under the head of "Shock," p. 25. The person suffering from fainting s h o u l d b e p l a c e d o n t h e b a c k, i f p o s s i b l e, the head slightly raised, if at all, obstruction to the circulation in the shape of cravat and collar removed, and any obstacle to perfect movement of the chest likewise dispensed with. For an adult, a teaspoonful of Brandy, in a little water, may be given every few minutes, until consciousness and restored action of the heart is observed. Twenty drops of Aromatic Spirits of Ammonia, in a teaspoonful of water, at short intervals, say every five or ten minutes, is quite as useful, but not always as easily secured. *Too much* stimulation in such a case might do harm, by causing the heart to send the blood with such force as to disengage the little clots spoken of at the divided extremity of the vessel.

If the loss of blood has been great, or the condition of the patient before the receipt of the injury such that the loss cannot be rapidly restored, the fainting may not rapidly or completely disappear. In such cases, beef tea and easily digested nutritious food, and even tonics, will probably be recommended by the medical attendant.

F o r e i g n   m a t t e r s, such as have been introduced into the wound at the time of the injury or subsequent to it, of course, should be carefully removed.

Having thus referred to certain features *common* to most wounds, the special, and what may be called the *distinctive* points of each class, according to the arrangement herein adopted, will now be given.

Incised Wounds.

After the h e m o r r h a g e ceases, and the c l o t s, with any f o r e i g n matter have been carefully and gently r e m o v e d, by a judiciously directed stream of water from a sponge, the separated s u r f a c e s a n d e d g e s o f t h e w o u n d s h o u l d b e b r o u g h t c a r e f u l l y t o g e t h e r. To *retain* them in position until union has taken place, strips of adhesive plaster may be used. This being a resinous preparation, it soon becomes dry, and useless for the purpose. Hence get but little at a time, and replenish with recently prepared as often as is necessary. In cities it can usually be had good from the large shops, where large sales prevent an accumulation of stock. With a pair of scissors, cut it lengthwise, into uniform sized strips of about a quarter of an inch in width, or even less in some instances. These can be subdivided in length, so as to extend across the wound, and far enough on each side to secure a suitable *hold* on the skin. Warm the plaster side of the strip at the fire until it becomes thoroughly and uniformly melted, then beginning with one end (recollecting that the *center* of the strip should cross the incision) rapidly and completely attach it to the skin, as a rule at *right angles* to the line of the cut. As the middle part approaches the wound, with the fingers bring up the skin towards the incision, from the *other* side, upon which the other half of the strip is to rest; then rapidly attach the rest of the strip.

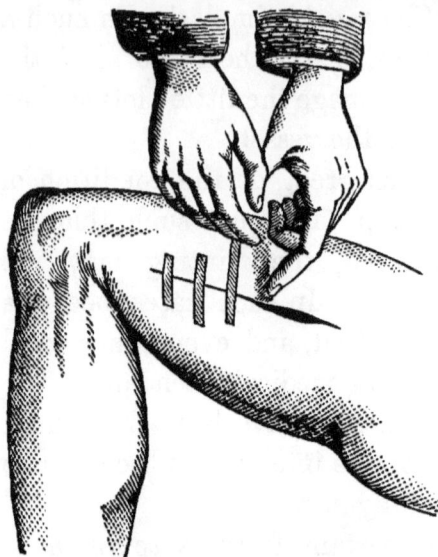

If this is not done, the strip of plaster will be found in folds, owing to the yielding of the soft skin beneath, and the edges of the wound *separated*. If *one* strip will keep the edges approximated along the whole length of the wound, no more is needed. If not, use others. Where more than one is used, the edge of the strip should be brought across a s h o r t d i s t a n c e f r o m t h e e x-t r e m i t y o f t h e w o u n d, so as to permit the ready exit of blood or pus. Confinement of either or both by the plaster, or anything else, favors " burrowing," as it is called, and consequent *separation* of the wounded surfaces. On the scalp, the face of men, and the extremities of some persons, the hairs must first be shaved off the skin, or the plaster will not remain attached.

Most persons, in using adhesive plaster on a wound,

apply a *large piece,* or several small pieces, so as to completely *cover* it. This must not be done. A few drops of blood escaping after *such* an arrangement, even when the edges of the wound have been carefully brought together, undergoes decomposition, irritates and inflames the parts, loosens the plaster, and *changes* what otherwise should have been the result of the accident.

Adhesive plaster is best for use, but as it is not portable, often in an emergency it is easier to get *Isinglass* plaster. This is a thin tissue of silk spread on one side with a solution of isinglass or other gelatinous substance. *Heat* is necessary to soften *adhesive* plaster; but *moisture* dissolves *Isinglass* plaster. As in most wounds there will be some liquid discharged, it can be seen at a moment's thought why Adhesive plaster will *remain* attached, while Isinglass plaster will become detached.

When Isinglass plaster only can be procured, it should be cut into narrow strips, the adhesive side moistened, and then applied as above directed. The black variety of Isinglass plaster, usually sold in small envelopes, is scarcely fit, as a general rule, to be used.

Speaking of plasters, the writer will say that on one occasion, where none was to be had, the edges of an incised wound of some length were successfully brought together, and held there, by a *postage stamp* divided lengthwise into four strips.

In shaving the face cuts are sometimes made which bleed to a troublesome extent. A crystal of common alum should always be kept with the apparatus, the bleeding absorbed by a fold of the towel, and then, before the blood can accumulate, thrust into the incision the edge of the crystal, holding it there a few minutes. If the bleeding continues, it is because the alum does not reach the divided vessel, and the wound should be wiped out until it can.

If the incision is deep, or there are not good points for

4

attachment of the plaster, *sutures* are often employed by
the surgeon. No definite rule can be given for the circumstances *requiring* them or the method of *using* them. An
ordinary sewing needle will not answer, but a needle with
a *cutting edge*, such as saddlers and glovers use for stitch-
ing leather, can be secured, if the regular surgeon's needle
cannot be procured. The suture is of white silk, or possibly
white flaxen thread might answer in an emergency. Surgeons now generally use fine wire of silver or iron, as
metal irritates the part it comes in contact with less than
a rough thread.

The edges of the wound-having now been properly
brought together, and retained there, the next thing is
what is called the "dressing." All manner of things were
once used for this purpose under the impression that they
were healing. They are now used by surgeons simply for
protective purposes. The *simplest* are therefore the *best*.
Hence water is now used, under the name of *Water
Dressing*. As Isinglass plaster is softened by moisture,
water cannot, of course, be employed when this material
has been used for retaining purposes.

Take two or three thicknesses of what is called Patent
Lint, if it can be conveniently had, if not, of old linen, or
even old muslin, somewhat larger than the wound.* With
a pair of scissors or a sharp-pointed knife perforate the
folds, dip in cold water, and after squeezing out the excess,
evenly apply to the wound. To *retain* in position, a strip
or two of adhesive plaster can be thrown over, or a small
roller (bandage) may be lightly applied. Keep the linen,
or substitute, constantly *wet*, not moist, with water. Sometimes the wounded member is supported in a sling.

* By "old linen," many persons think the linen bosoms of old shirts is
meant. For the purpose mentioned it is practically useless. An old
damask linen table-cloth furnishes the best, and next to it, perhaps, old
linen sheeting, quite coarse in texture.

This dressing is so simple, and at the same time so useful, that surgeons are apt to use no other in simple wounds; but, unfortunately, it is so simple that many persons, unless they are intelligent, have no confidence in it. They prefer pain-killers, liniments, herbs, and salves. Remember, the *natural* reparative process unites the parts, and the effort of the surgeon is only to put the parts in the position best calculated to favor this to advantage. All foreign matters, never mind under what name; as a rule, are obstacles, not aids, to this process of nature.

In using water as a dressing, or applications of any kind to the surface, if a sense of chilliness appears, its use should be discontinued for a time. As said elsewhere, the conversion of the liquid to a vapor requires much more heat than might at first be supposed. When the chilliness is observed, a little of some kind of stimulant may often be useful.

If the pain is severe, sometimes opium, in the form of Tincture of Opium (Laudanum) is added to the water applied.

After a certain time, usually twenty-four hours, occasionally sooner, sometimes later, the outside strips of plaster holding down the *lint* should be divided, the parts removed, and the lint carefully removed, after loosening it as far as is practicable by moistening with tepid water. Should any portion be closely adherent to the wound, or any part of it, through coagulation of blood or escape of pus, and fail to become detached under delicate manipulation, with a sharp pair of scissors divide the lint as near as convenient to the point of adhesion, letting the fragment remain, with the hope that by the next time separation can be secured.

The adhesive plaster now only remains, and if it has been properly applied the condition of the wound can be easily determined. If there is no discharge of blood or other material, the plaster should be let alone, and another piece of lint and retaining strips applied, and kept wet with water as before.

The next day the same examination should be made. If blood or pus is found, remove it with a soft piece of moistened sponge, being very careful not to disturb the wound or the strips of adhesive plaster. Should any strip have become loosened, remove it by catching hold of the extreme end and separating it gently and slowly until detached almost up to the line of the incision; then drop that extremity, taking up the *other*, and go through with the same thing until only the central portion over the wound remains to be separated. This is done to lessen the chance of tearing the wound *apart*, which pulling at *one* end of the strip would favor. With a little soap and water then remove the remains of the resinous portion of the plaster from the skin, dry gently and well, and apply a fresh strip as a substitute for the old, observing all the precautions suggested.

These remarks apply, of course, to a simple incised wound, when union takes place at once, or with but little suppuration (making of pus). This cannot always be secured, from suppuration of the sides of the wound after the dressing has been applied, or an unfavorable condition, as it is said, of the blood. In such a case the blood or pus must be removed once a day, as a rule, the surfaces of the wound kept together, as much as possible, by adhesive strips, until a junction is effected. Do not use too much soap and water, as the only object of them is to better and more easily remove the foreign matters (blood and pus), which, if retained, act as irritants; but not to remove the *reparative* material poured out by nature for *joining* the separated surfaces.

If, owing to the general health of the patient, or a new character given the wound by some unavoidable mishap during the course of treatment, there should be decided suppuration, the injury may require more frequent dressing, especially in hot weather. In such a case, if the wound has up to this point been without professional advice, it may be better to consult a physician, as a suitable tonic, a different diet, or even some local applications to the seat of injury, may be followed at once by an improvement in its appearance.

Under the classification adopted, have been included with incised wounds those instances where portions of the body have been cleanly *cut off*. Never mind what part it is, if the excision has recently taken place, t h e s e p a r a t e d p o r t i o n s h o u l d b e t a k e n, rapidly freed from any foreign matter, a n d a p p l i e d t o t h e p a r t f r o m w h i c h i t h a s b e e n s e p a r a t e d, i n t h e p o s i t i o n i t p r e v i o u s l y o c c u p i e d. Should the weather be cold, some raw cotton might be applied around it to preserve the *warmth*, and some measures inaugurated by which gentle and uniform *pressure* can be kept up for a reasonable time. After making all allowances for the remarkable stories told in reference to such things, there is no doubt that much can be said in favor of the practice and little against it; for if circulation and adhesion are not restored, it can be said that only a little time has been lost.

Punctured Wounds.

These vary in their importance, not only according to the *depth* of the wound and the *structures* penetrated, but according to the *instrument* inflicting them. The chief peculiarity and danger of these wounds is, that their nature does not afford *sufficient* facility for the escape of

blood, other fluids, or foreign matters. The retained fluids
decompose, or, by mere pressure, irritate the adjacent
parts, or, by distention, enlarge the original wound.

In *punctured* wounds the essential idea is to treat it as
an *incised* wound; but its peculiar character, greater in
depth than external area, requires a somewhat different
plan of procedure. R e m o v e whatever f o r e i g n m a t-
t e r s have entered the wound, and apply a pad to the
outer wound, so that it will, if possible, moderately and uniformly exert some pressure along the deeper portion. The
wound fills up with blood, and as it cannot escape externally, on account of the pad, it clots, and closes the open
ends of the divided vessels, by *pressing* upon them. If the
wound was made with a *blunt*-pointed instrument, it is practically, as far as *union* of the divided surfaces is concerned,
a *contused* wound, which will next be alluded to, and will
heal as such. If with a *sharp* instrument, the wound often
heals as an incised wound; and if there is no discharge to
require it, the pad may be left in position, as strips of adhesive plaster would be in an incised wound, without disturbance, until union of the divided surfaces is complete.

In case much pain follows, with signs of inflammation
around the injury, the dressing (pad) must be removed, to
permit the *escape* of the results of inflammation of the
deeper portions of the wound. Sometimes even the *external* opening of the original puncture is not large *enough*
for the exit of pus and other discharges. In such a case
the surgeon must enlarge it until the requirements in this
respect are properly met.

Once a day, or oftener if the wound is discharging, it
should have the dressing changed, to insure neatness and
escape of pus. If certain structures are invaded by the
puncture, the surgeon is often at a great deal of trouble to
insure healing at the *bottom* of the wound first, to guard

against the *burrowing*, as it is called, of pus between the muscles and other contiguous parts.

Under the head of "Punctured Wounds," may be mentioned a trivial set of injuries, quite frequent in occurrence and often attended with serious *consequences*. They are produced by the running in of a thorn, splinter of wood, or a piece of metal. The foreign body is pulled away in most cases, if it can be done readily; if it cannot, it is let alone, as the phrase is, "to work out." In all cases, if a splinter or thorn, it should be got out. Not by poking at it with a needle, or something of the kind, which *adds* to the irritation, but by making an incision along its course, so as to expose it enough to get a sufficient *hold* upon it. If the incision should not permit a removal, a more ready escape has been made for the foreign body and any pus ("matter") that may form; thus lessening the probability of the constitutional excitement exerted through the nervous system known as Tetanus (Lock-jaw). If the splinter is under the finger nail, and cannot be pulled out, do not waste the outside end by picking at it. The nail immediately *above* should be *scraped* as thin as possible by a piece of glass, and then the thin nail overlying should be split with the blade of a knife, or an incision made on each *side* of the splinter, the tongue of nail between the incision removed, which should expose the upper surface of the splinter along its entire course. The restraining pressure of the nail upon the foreign body is in this way gotten rid of, and at the same time an outlet for the products of inflammation is given.

A piece of lint, wet in water, to which a good deal of Laudanum has been added, should be applied, and kept wet with it as long as may be necessary.

When the finger or hand, toe or foot, has been pricked, particularly by anything foul, as a rusty knife or nail, a dirty piece of horn, or bone, the opening does not permit the escape of the retained foreign particles, and inflammation results. The skin on these parts is so thick that it cannot yield when the parts beneath are irritated and inflamed, and the inflamed portion, as it were, tightly bound up, or squeezed as in a vice, by the hard skin, and the almost always fatal condition of affairs known as Tetanus (Lock-jaw) supervenes in many cases.

Whenever such wounds, to such parts, are received, a n i n c i s i o n s h o u l d b e m a d e i n t o t h e p u n c-t u r e , thereby providing a suitable escape for the blood, pus, etc.; and a piece of linen dipped in Laudanum forced into the wound. This can be done by almost any one, and may save serious trouble.

In washing clothing, scrubbing and scouring, a fragment, or even an entire needle is sometimes forced beneath the skin. Do not attempt to get it out, but hold the part perfectly quiet until a surgeon can be procured. The slightest movement often places it beyond detection of the sight or touch. When this happens, there is no occasion to be alarmed, as the *needle* slips in between the muscles, and cannot even be felt as painful. It does no harm there, as inflammation almost never results. Occasionally it is unexpectedly found near where it entered, and in a position favorable for extraction.

Contused Wounds.

As the name implies, these are divisions of the tissue with contusion (bruising) of the parts. Some of the tissue is generally removed, the edges rough and irregular, but there is generally less gaping of the edges and less bleeding, than of *incised* wounds of the same extent. The contusion impairs the contractility, as it is termed, of the parts, hence the less gaping of the edges; and the blood vessels have been *torn,* and the roughened extremity of each vessel soon favors a clot there, hence the less *bleeding.*

Contused wounds *especially* need careful cleaning out and removal of clots. The *general* procedure of treatment may be the same as for Incised Wounds, but with more watching for the occurrence of deep-seated inflammation and sloughing away of the contused edges and surfaces during the later process of suppuration. As soon as any alarming bleeding has been checked by the application of ice or cold water to the blood vessels, or, if necessary, by pressure upon them, bring the edges of the wound together by strips of adhesive plaster; *remembering* in applying them to a contused wound, that there must necessarily be inflammation of the bruised parts, with consequent discharges. It is rarely that the *entire* wound is contused, although that may be its *general* character; so a portion of it, often the extremities, unite as an *incision,* leaving the *rest* of it to pursue a different course. After sloughing (separation of the part whose vitality has been destroyed by the contusion from the living portion) has begun, *poultices* are often of use in favoring the process. After due time the parts damaged beyond repair become detached, and the contused wound appears as a cavity, more or less superficial in depth, lined with a velvety surface, more or less obscured with pus. After a while the gap is *filled up* by these granulations (new, but immature flesh). When these granulations get above the surrounding edges of the wound, before they become consolidated, the appearance is popularly termed "proud flesh," and wrongly supposed to prevent the healing of the wound.

These granulating surfaces are sometimes stimulated to increased activity by the application of some simple stimulating ointment, or by gently brushing once a day, by means of a camel's hair pencil, or the floating edge of a clean, soft feather, with a solution of Sulphate of Copper (blue vitriol), say a piece as large as a grain of coffee dissolved in a couple of tablespoonfuls of water.

Lacerated Wounds.

Lacerated Wounds are made by rending or *tearing* the parts, rather than by cutting, as in Incised, or by breaking, as in Contused Wounds. The treatment follows that of the latter named injuries, to which they bear a strong resemblance in most respects. The chances for union are subject to about the same probabilities, and should be favored the same way. These wounds often occur in the scalp; sometimes a large piece is detached, and left hanging by a small attachment. *Never,* under any circumstances, permit the fragment to be *removed,* as the scalp is so largely supplied with blood-vessels, that injuries there of the most unfavorable aspect to an ordinary observer are often rapidly and completely repaired.

Poisoned Wounds.

This includes all wounds into which any poison, venom, or virus is introduced. The wound is not always made at the time of the introduction of the poison, but often exists previously. Thus, a scratch, fissure, or ulcer may exist on the hand or other part, and afford entrance to alkalies, acids, or other such irritants. When such a wound of the skin exists, and may be exposed to any irritating substance, great care should be exercised to prevent trouble arising. Persons with such injuries to the hand have suffered most severely from skinning animals which have died of pleuro-pneumonia and other diseases. Physicians often decline making a *post-mortem* examination because a wound exists on the hand, and many have died under the circumstances, because they did not abstain.

If a poison should be introduced, and at once observed, before *absorption* by the system has taken place, a stick of Nitrate of Silver (lunar caustic) should be thrust into the wound, or, what is more certain would be to heat a large nail red hot and force the end into the opening.

Another variety of poisoned wound is when the poison is introduced at the time of the injury, as in the case of b i t e s   a n d   s t i n g s   o f   i n s e c t s ,   and the   b i t e s of s e r p e n t s and a n i m a l s .

In this latitude there are but few i n s e c t s known whose bites can really be considered poisonous. A swollen face which cannot otherwise be accounted for is often attributed to the bite of a S p i d e r . Trouble rarely results from it, and when it does it will often be safe to ascribe a portion to the condition of the general health at the time. The bites of certain f l i e s have been followed by symptoms of local poisoning, but it would be well, should it happen, to know where the fly had been just before.

The s t i n g s o f i n s e c t s , as Hornets and Bees, are always painful, and sometimes followed by great swelling. A stimulating application to the injury, as a drop of Aromatic Spirits of Ammonia, will often afford the greatest relief. A pinch of common table salt, dampened with water and rubbed in, is very useful for the same purpose; likewise a slice of onion rubbed on gives almost instant comfort.

There are few s e r p e n t s , likewise, in this latitude, whose bite is followed by poisonous symptoms. The bite of the common Rattlesnake is a well known exception to the rule. Where a person has been bitten by one of these serpents there is usually little time to be lost. If it is the hand or other accessible part, the fold of the skin containing the puncture should be gathered between the teeth, and the strongest suction of the lips used to extract the venom. If there is no laceration or other injury to the skin of the lips, or the mucous membrane of the mouth, it can be done with *impunity*.

The symptoms are a slackened action of the heart, indicated by a feeble pulse and other appearances of prostration, indicating the free use of stimulants. Marvelous stories are told of the quantities of whiskey and brandy taken under these circumstances by persons not addicted to the use of them. Either are usually to be had on such occasions, and it might be wise to give of them freely at brief intervals, until symptoms of slight intoxication appear. As the heart is much enfeebled in its power, it would easily suggest that the person bitten should be made to lie down on his back, as that is the position where the strength of the heart is least taxed.

BITES.

Independent of the consideration whether any *poison* has been introduced through the wound, Bites may be regarded as a *lacerated* as well as a *contused* wound. There is usually a good deal of sloughing of the bitten parts, and no small amount of pain, owing to the nature of the wound. Care should be taken to remove from the wound any particles of clothing, should any have been forced into it, then wash out with tepid water with a little Castile soap. Usually the part is so much *contused* that no effort is made to secure adhesion of the opposite sides of the wound; but water dressing is at once applied, and suppuration and sloughing awaited. The pain is often quite severe, in which case some anodyne, in the shape of Laudanum, should be added to the water used as the dressing.

Bites of Dogs.

Rabid dogs are much less frequent, perhaps, than is generally thought; and a rabid dog, it may be supposed, might bite many human beings without necessarily communicating Hydrophobia. In the first place, the chances are that the saliva would be *arrested* by the *fabric* over the part bitten, if there should be saliva in the mouth of the dog at the instant; and if, as is said, the saliva is not *itself* poisonous, but that the poison is *mixed* with it, the saliva might not, at *that* moment, contain any. It is stated, by what is considered competent authority, that of dogs bitten by others known to be hydrophobic, scarcely more than one in four become affected; and it is likewise said, that among human beings, when no precautions are taken, not more than one in ten or fifteen are affected after being bitten. The celebrated surgeon, John Hunter, knew of twenty-one people who were bitten by the same dog, and only one of the number had the hydrophobia. It should be added, however, that it is not stated that this individual had not been bitten by some other dog than the one which bit him in common with the rest. Besides, many persons have, undoubtedly, died, after having been bitten, with convulsions, not of hydrophobia, but the result of anxiety and fright. One well-known physician, after having been bitten, as a precautionary measure, blew out his brains.

Some writers, of no mean repute, assert that the bite of a healthy dog, when under a state of anger or fright, may communicate hydrophobia, or another disease like it, from some change effected by the emotion, in the character of the saliva. It is likewise contended that it may spontaneously arise in animals.

However, as these things cannot be demonstrated to the

satisfaction of the victim or his friends, and there is no
known remedy for the disease, it is a l w a y s b e s t,
a f t e r a b i t e b y a s u s p e c t e d d o g, to a c t
"o n t h e s a f e s i d e."

Therefore, *at once*, remove the clothing, if any, from
the bitten part, and apply a temporary ligature *above* the
wound. This *interrupts* the activity of the circulation of
the part, and to *that* extent delays the *absorption* of the
poisonous saliva by the severed blood-vessels of the wound.
While other things are being hurriedly prepared for, some
one whose lips and mouth are free from breaks might
attempt s u c t i o n o f t h e w o u n d. The material
extracted by the act, apparently chiefly of blood, should,
of course, at once be ejected from the mouth of the person
giving the assistance. The *bite* is *really* a *lacerated* and
contused wound, and lying in the little roughnesses, and
between the shreds, is this poisonous saliva. If by any
means these projections and depressions affording the lodg-
ment can be removed, the poison must go with them. If
done with a knife, the wound would be converted, practically, into an incised wound, and would require treatment
as such. If a Surgeon is about, he would probably stand
a probe upright in the wound, and with a sharp knife
c u t t h e e n t i r e i n j u r e d p o r t i o n o u t. Professional aid is not always at command, and in such a case
it would be well to take a poker, or other suitable piece
of iron, heat it *red* hot, at least, in the fire, wipe off and
d e s t r o y t h e e n t i r e s u r f a c e o f t h e w o u n d.
As fast as destroyed, the tissue becomes white. An iron
at *white* heat gives less pain than one "black hot," as
smiths say; for in the latter instance the heat is scarcely
sufficient to *destroy*, but only *irritates;* while in the former, the greater heat at once destroys the vitality (kills)
of the part with which it comes in contact. With a pro-

perly heated iron, not only the *surface* is destroyed, but the destructive influence extends beyond and into the *healthy* tissue, far enough, if no point is neglected, to assure the purposes for which it is used.

Some are inclined to think that if the wound is at once well wiped out, and a stick of solid nitrate of silver (lunar caustic) rapidly applied to the entire surface of the wound, that little danger is to be apprehended. It acts, but in a milder degree, like the heat of the iron upon the tissues. In case the heat or the caustic have been used, poultices and warm fomentations should be applied to the injury, to hasten the *sloughing* away of the part whose vitality has been, in this instance, *intentionally* destroyed.

There is a strange belief among the ignorant, particularly among the people from Ireland, that, whether the dog was "mad" or not at the time of giving the bite, if it should become so at any *future* time, the disease will appear in whatever individual the animal has bitten. A dog after having bitten a person, is apt, under this mistaken belief, to be at once slain. This should not be done, but the *suspected animal placed in confinement*, and watched, under proper safeguards, for the appearance of the disease. Should no satisfactory appearances indicate the disease in the dog, it can be seen, in a moment, what unnecessary mental distress can be saved the person bitten and his friends.

This mysterious disease, although known from the days of Homer and Aristotle, has never yet been cured or understood. Animals communicate it to each other, and to men, by the bite; but no known instance is recorded where one human being has communicated the disease to another, although many patients, in their spasms of Hydrophobia, have bitten their attendants, as they have done in spasms from other causes.

There are many popular errors in reference to this
disease, some of them most grotesque in character. This
terrible malady is known among scientific men as Rabies
Canina (Rage of Dogs); but, from one of its symptoms,
Hydrophobia (Fear of Water). So far from *fearing*
water, the poor animal seeks it; but, owing to a spasm of
the muscles of the throat, it is unable to quench its terrible thirst. Another prevalent but erroneous belief is,
that the disease prevails among animals in the hot weather
of midsummer, while the truth is, that it is more apt to
occur in Winter, or the damp, cold days of Spring. As
so little is known of the disease in the dog, and another
common disease of the same animal (distemper) is often
associated with it, the following, from Youatt, is inserted.

Mr. Youatt, whose description of Canine Madness is generally quoted
and accepted, says: "The disease manifests itself under two forms: the
furious form, characterized by augmented activity of the sensorial and
locomotive systems, a disposition to bite, and a continued peculiar bark.
The animal becomes altered in habits and disposition, has an inclination to
lick or carry inedible substances, is restless, and snaps in the air; but is
still obedient and attached. Soon there is a loss of appetite, and thirst;
the mouth and tongue swollen; the eyes red, dull, and half-closed; the
skin of the forehead wrinkled; the coat rough and staring; the gait un-
steady and staggering; there is a periodic disposition to bite; the animal
in approaching is often quiet and friendly, and then snaps; latterly, there
is paralysis of the extremities; the breathing and deglutition become
affected by spasms; the external surface irritable, and the sensorial func-
tions increased in activity, and perverted; convulsions may occur. These
symptoms are paroxysmal, they remit and intermit, and are often excited
by sight, hearing, or touch.
"The *sullen* form is characterized by shyness and depression, in which
there is no disposition to bite, and no fear of fluids. The dog appears to
be unusually quiet, is melancholy, and has depression of spirits; although
he has no fear of water, he does not drink. (The fear of water, it should
be said, is acquired by experience, the effort of swallowing being attended
with spasm of the muscles of the throat, afterwards often extending to the
rest of the muscles of the body.)
"He makes no attempt to bite, and seems haggard and suspicious, avoid-
ing society, and refusing food. The breathing is labored, and the bark is
harsh, rough, and altered in tone; the mouth is open from the dropping

of the jaw; the tongue protrudes, and the saliva is constantly flowing. The breathing soon becomes more difficult and laborious; there are tremors, and vomiting, and convulsions."

In a recent paper* the writer refers to a large number of well-authenticated instances where the bite of the common Skunk, or Polecat (Mephitis mephitica) has been followed, after the usual period of incubation, by symptoms of Rabies (Hydrophobia). Of the forty-one cases mentioned, every instance but one (a farmer, who knew of the danger, and had taken the precaution of using prompt preventive treatment) ended in death. This is more fatal than the bite of the rabid dog.

The wide distribution of this animal, the common Skunk, over the United States, and the readiness with which people might be exposed to its bite, should lead persons so injured by it to at once resort to the peculiar measures advised for the treatment of bites of suspected dogs.

FOREIGN BODIES IN THE EYE.

Particles of cinder, dust, or fragments of metal, often get into the eye, and cause a good deal of trouble. Sometimes they are dislodged, and washed out by the extra secretion of tears brought about by the irritation produced by the body. Sometimes this process does not give relief, and it is necessary to resort to some p r o c e s s o f e x-t r a c t i o n. A popular, and often useful plan is to take hold of the lashes of the upper lid, separate it from the eyeball, so that the lashes of the lower lid will slip up in the space, acting as a brush to the inner surface of the upper eyelid. This, of course, cannot remove anything,

---

* Rabies Mephitica. Hervey, Rev. H. C. American Journal of the Sciences and Arts (Silliman's) May, 1874, p. 477.

as a rule, from the eyeball.   A better way is the usual one
of holding a knitting-needle over
the upper lid, close to and just
under the edge of the orbit,
then, holding it firmly, seize the
lashes of that lid by the fingers
of the disengaged hand, and
gently turn the lid upward and
backward over the needle, or
substitute used.   Movement of
the eyeball by the sufferer, in a strong light, usually re-
veals the presence of the intruding body, so that by means
of a corner of a silk or cambric handkerchief, it can be
detached and removed.

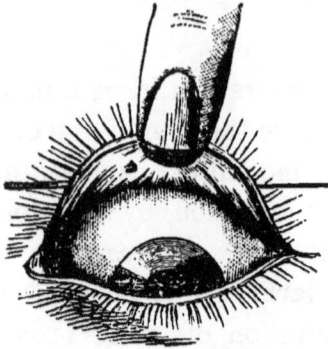

Should the foreign body be *imbedded* in the mucous
membrane covering the eyeball or the eyelid (conjunctiva),
a steady hand and a sharp-pointed instrument will usually
lift it out.

The foreign body often cannot be seen, but the person
assures us that he *feels* it.   Often he does not really feel
the *presence* of the body, as much as the roughness (really,
a wound) left by it.   In such a case, or even if the body
has been seen and removed, a soothing application to the
injury is as useful as the same thing applied to a wound of
the hand.   Take a spoon or cup, heat it, and pour in a few
drops of Laudanum.   It will soon become dense and jelly-
like.   A few drops of water added will dissolve this gummy
material, and the liquid thus formed may be applied by the
finger to the "inside of the eye," as they say.   The Laudanum is Opium dissolved in Alcohol.   The Alcohol is
somewhat irritating, but is easily *evaporated* by the gentle
heat, leaving an Extract of Opium, which is dissolved in
the water afterwards added.

The comfort derived from this simple and always accessible preparation, after injury to the eye by a foreign body getting into it, is of the most satisfactory kind. In no case use any of the popular " Eye Waters," or " Salves."

Not an uncommon accident is a fragment of l i m e i n t h e e y e. The delicacy of the organ, and the activity of this powerful Alkali, require all that is to be done to be done at once. Do not waste time by attempting to *pick* it out, but *neutralize* the alkali by a f e w d r o p s o f V i n e g a r (which is dilute Acetic Acid) in a little water. A few drops of Lemon Juice, in a little water, will answer just as well, if introduced, like the vinegar, into contact with the lime. Even when done rapidly, the ulceration caused by the Alkali will be some days in disappearing. In all cases where lime has entered the eye, even when these things have been used, no time should be lost in going to a Surgeon.

FOREIGN. BODIES IN NOSTRILS AND EAR.

The curious disposition of children to insert foreign bodies, as grains of coffee, corn, peas, pebbles, etc., up the nostrils, and into the ear, is too well known to be more than alluded to. If the body is *soft*, it absorbs moisture from adjacent parts, becomes *swollen*, and more difficult to remove. If the body is *hard*, the irritation and *inflammation* set up by it in contiguous parts in a short time materially increases the difficulties of removal. Hence the *sooner* these substances are removed, the easier it is to do so.

Foreign Body up the Nostril.

If the foreign body is up the nostril, the child should be made to take a full inspiration (" a full breath"), then closing the other nostril with the finger, and the mouth with the hand, the air of the lungs, in escaping through the nostril closed to a degree by the foreign body, assisted by a sharp blow from the palm of the hand to the back, will often expel the substance.

If it will not escape in this way, and it is near the opening of the nostril, compression by the fingers, just above, will prevent it getting further up, and it can be hooked out with the bent end of a wire or bodkin. Should these measures not remove the foreign body, the child should at once be taken to a Surgeon.

Foreign Bodies in the Ear.

Foreign bodies in the ear are more troublesome to deal with. No effort to remove them with a *probe,* or anything of the kind, should be made by *any* one, except a professional man, for fear of *permanent injury* to the ear. The head of the child, face downward, should be held firmly between the knees, and with a Mattson's or Davidson Syringe a stream of tepid water should be injected into the ear. The nozzle of the syringe should not be introduced into the *cavity,* as its presence may *prevent* the dropping out of the desired body after the water has been forced past and beyond it.

Should this means not succeed, consult a Surgeon without delay.

I n s e c t s sometimes get into the ear. The best way of getting them out is to hold the head of the person with the disabled ear upward, and fill the cavity with sweet oil or glycerine. It drowns the animal, by closing up its breathing pores, and in a short time it floats to the surface of the fluid used. The tube of the ear is somewhat curved, and when straightened somewhat by catching hold of the upper tip, and gently pulling it upward toward the crown of the head, the liquid flows in more readily.

FROST BITE.

Exposure to the cold, of severe degree, often leaves the fingers and toes, nose, ears, and lips, more or less frozen. This condition, short of absolute *death* of the part, is termed Frost Bite. It will be observed that the portions of the body just enumerated are those most exposed, in area, to the influence of the cold, and are furthest situated from the heart; and it will, perhaps, be unnecessary to remark that persons who are *debilitated* are more apt to suffer with the same amount of exposure than the *robust*.

When the circulation of any part begins to succumb to the influence of the cold, it becomes puffy, blueish, and smarting. This is because the blood moves more slowly than natural through the vessels exposed near the surface. Soon this blueness disappears, and the part becomes pallid, as if the influence of the cold had contracted the vessels to an extent incompatible with the passage of blood through them. The *pain* at this point ceases; indeed, until he meets a friend, he often does not know of his mishap. At this stage the injury has become so great that, unless proper means are taken to restore circulation, complete *death* of the part ensues, and in due time sloughs away, and is detached from the line of living tissue.

What takes place in a *part* of the body, known as Frost Bite, may take place in the *whole* of it, which is known as "Frozen to Death." The blood of the extremities being gradually forced from them, under the continued subjection to the cold, is forced inward upon the larger blood vessels, heart, lungs, and brain. There is increasing difficulty in breathing, owing to the engorged state of the chest, and, what should always be remembered by one so exposed to cold, an *unconquerable desire to sleep*. To sleep *then* is to die. If the person exhibits such a symptom, he must, by all means, be kept constantly moving.

Treatment.

Persons exposed like those just described must be treated promptly, and with one thing never lost sight of. That is, keep the frozen person away from the heat. A person taken up insensible, or approaching it, from exposure to the cold, should be taken into a *cold* room, his clothing removed, and thoroughly rubbed with snow, or cloths wrung out with ice water. The friction to every part of the body, particularly the extremities, must be continued for some time, until signs of returning animation appear. When the frozen limbs show signs of life, the person should be carefully dried; put in a cold bed in a cold room; Artificial Respiration used until the natural is established; then brandy given, also ginger tea, and beef tea. Usually, by this time medical advice will have been secured to direct further treatment. Should it not, do not forget that the patient is to be brought by degrees into rather warmer air; and lest in some *part* there might still be defective circulation, the person should be kept away from exposure to the heat of the fire.

Milder degrees of the same condition, as suspension of life in the ear, nose, finger, or toe, from exposure to cold, must be treated with the same general directions in view. The part should be kept away from the heat, and rubbed with handfuls of snow, or towels dipped in cold water, until circulation appears re-established. Exposure of the part to the heat before, we may say, it has been almost *rebuilt*, is apt to be followed by *sloughing*.

CHILBLAIN,

As the name implies, is when the circulation of the part has become chilled—*disturbed*, not destroyed. It is generally attended with much itching, tingling, and smarting, and is usually found in the toes, outside edge of the feet, just where the toe emerges, or in the heel. Sometimes, in persons of debilitated state of health, the hands suffer. These symptoms are particularly annoying just after lying down in bed, owing to the exposure to the heated air formed and retained between the bed-clothing by the body.

The most useful thing for these annoying symptoms, and a condition which often extends into ulceration and sloughing, is to keep away from the fire, and every night, before retiring, to bathe the feet in cold water, or rub them with *snow*. They should then be well dried with a soft towel: After this, the application of the ordinary Compound Resin Ointment of the apothecaries is often of use in stimulating the circulation through the part. The efficiency of this ointment for this special purpose can be increased by asking the apothecary to add to an ounce of it a couple of drachms of the Oil of Turpentine. It may be remarked, that persons who suffer in winter from cold feet are often benefited to a surprising degree by bathing them at night, before retiring, in *cold* water. Such persons should always keep their feet away from the fire.

CONVULSIONS.

Convulsions, or "fits," as they are often called, are a frequent cause of alarm in the streets, or at public assemblages. In the decided majority of instances, the convulsions may be safely presumed as Epileptic; so, unless otherwise specified, the remarks here made apply to that form. Ordinary *fainting* may be confounded with it; but here the face is pale, the person perfectly still, and there is no perceptible breathing. Besides, in fainting there are no *convulsive* movements.

Often the Epileptic seizure is ushered in with a peculiar *sharp cry*, as the person falls over. It does not always occur, but when it *does* there can be no doubt, if it is a convulsion at all, that it is Epileptic. There is frothing of the mouth, sometimes tinged with blood from the tongue or a fold of the lips having been caught between the teeth at the moment the spasm commenced in the muscles of the jaws. Sometimes there are general convulsive movements of the whole body; often of parts of it only. At first the face is pale, but usually, in the course of a few moments, it becomes livid, except around the mouth, which often continues pale, in strong contrast with the color of the rest of the face. As a general rule, it may be said that the convulsive feature of attack does not last much longer than four or five minutes, although to bystanders the time naturally seems longer. Then the person opens his eyes with a certain degree of intelligence, or revives enough to speak; and, as will be said, it is at this point of the attack that most must be done. Sometimes there is nothing beyond it, and the individual gets up, hurriedly puts on his hat, and walks off, apparently the least concerned of anybody about.

If this happy termination does not take place, this brief semi-conscious interval gives way to a *heavy stupor*, varying in duration from thirty minutes to three or even six hours.

In Epileptic Convulsions there is usually nothing to be done. Ignorant people on such occasions are apt, upon the general plea, "if you do not know what to do, do something," to insist upon "opening the hands," as the phrase is, saying that the patient will be better as soon as they can do it. The truth is, they cannot do it until the patient is better. All interference of this kind is *hurtful*, and no good can come of it. All rude efforts aggravate the trouble, perhaps, by exhausting still further the muscular strength of the patient.

All that can be done is to keep the person from injuring himself or hurting others during the violent convulsive movements, by r e m o v i n g h i m t o s o m e c l e a r s p a c e w h e r e t h e r e i s n o t h i n g t o s t r i k e a g a i n s t. Do not attempt even to hold the limbs, but loosen everything about the throat and chest.

Treatment.

Wait a few minutes for the convulsive movements to cease, and the semi-conscious state to appear. As said above, it will soon be seen. Then, if the person is a stranger, get his *name* and *residence*, if possible, with such other knowledge as may be useful. In the meanwhile, keep the crowd away. This is a very important measure of assistance in convulsions, as in every other emergency. By this is not meant so that people cannot bend over the victim, but that a *perfectly* free space of at least two feet on each side should be kept, with none in it but the one or two immediately assisting him.

Thirty drops of the Aromatic Spirits of Ammonia, in a teaspoonful of water, may be given the patient, as it is thought by many physicians to lighten and shorten the later stupid stage. The spasmodic condition of the muscles of the jaws can usually be overcome enough, with a little gentle dexterity, to permit it to be got into the mouth with the assistance of another spoon or a piece of smooth stick. After getting the liquid into the mouth, press down the base of the tongue, and the mixture will readily run down the throat. As much of it will necessarily be lost during the operation, double the quantity may be prepared for use. If more than the thirty drops should be given, no trouble from it need be feared.

If the name and residence have been secured, as it often can, at the interval alluded to, the friends of the person can be advised. If not, he should be taken to some place of security until consciousness returns.

Persons liable to Epileptic Convulsions should *never* be permitted to go from the house without a strip containing the name, residence, and disease, attached inside of the coat, where it will at once be seen upon unbuttoning the coat over the chest. A reference on it to a memorandum in some pocket containing a suggestion as to the duration of the attack, useful remedy, if any, in assisting restoration, would often materially add to the comfort and advantage of the afflicted person.

Other Convulsions are Apoplectic. These are not common, in comparison with others. As a rule, little can be done by bystanders, further than loosening everything about the neck. This should be done in all Convulsions.

The Convulsions known as Hysterical are usually found in young women who are not very strong. Until assistance comes, act as directed in Epileptic Convulsions. The distinction between them cannot be expressed, to a useful extent, to unprofessional persons.

POISONS.

Under this term people are inclined to place only those things which, if taken internally, produce death. Physicians, however, consider it merely a *relative* term, and call anything a Poison that does more harm than good to the body. A little of a good thing may be useful, but, beyond the point of usefulness, may be injurious. An exaggerated injury, from the same cause, may well be termed a poison. There is not a single poison in the entire list which, in proper quantities, and under favorable circumstances, may not be used with advantage to the human body; and, on the other hand, there is scarcely a single thing in ordinary use, which, if indulged in beyond the requirements of the body, or its ability to properly dispose of it, may not be followed by symptoms of derangement of the economy; and, in the above qualified sense, is not miscalled, if termed a Poison.

In the majority of cases, the poison is introduced into the body through the stomach. As soon as swallowed, a portion of the agent may commence *destructive* action upon the mouth, throat, or stomach, as in the case enumerated of Acids, Alkalies, Arsenic, Phosphorus, etc. While some substances act in this way, others pass from the stomach, through the mucous membrane, without injuring it, into the *blood*, and are carried by it to the brain and other portions of the nervous system, where the *really* injurious action begins, by overpowering them; so that the breathing and action of the heart are not kept up. To this class of poisons belong Alcohol, Aconite, Belladonna, Opium, Strychnia, etc.

A slight knowledge of the *mode* of action of a substance will, therefore, of itself suggest an antidote or remedy. If an Alkali has been taken, an Acid will *neutralize* it, converting it into a compound less hurtful. The new compound is, perhaps, *injurious*, but not so *active*, and can be removed from the stomach somewhat at leisure. On the other hand, if an Acid has been taken, an Alkali would naturally suggest itself as an antidote.

Some substances cannot be *neutralized* by any convenient article; the poison is then to be *removed* from its lodging place as soon as possible, and its effects *counteracted*.

If the agent does not act upon the stomach *directly*, but upon the brain and nervous system, reaching it *through* the blood, a recollection of what was said when certain gases have been inhaled will meet the case. Artificial respiration would, of course, be resorted to. This should continue until enough of the poison in the blood has been eliminated (thrown out) by the natural processes constantly going on *in* the body, until the brain and nervous system are able to resume one of their old duties, of attending to the respiration and circulation of the blood.

As few persons have the necessary knowledge of the different poisons, each of these substances will be spoken of somewhat in detail, and alphabetically arranged, so that, in case of need, immediate reference can be made to the particular substance supposed to have been taken.

Before saying anything further, it should never be lost sight of, that the substance swallowed as a poison must be considered as two parts; The portion of that taken which has *already* had an opportunity of acting upon the mucous membrane (lining) of the throat and stomach, if the poison acts in that way, or which has already passed from the stomach into the blood, if the poison acts in the other way; and the portion of the poison in the stomach *yet* to be disposed of.

It is the latter portion, perhaps, in most instances we are called upon to first deal with; and the means employed is, to evacuate the stomach with the least possible loss of time. This is done with the stomach pump, and by emetics.

Stomach Pump.

No directions for poisons are complete without reference to this piece of apparatus. With people who know nothing about the matter, it is very popular. The writer knows of but one physician among all his acquaintances who professes to keep one, and unless this particular instrument is different from all other complicated instruments rarely used, he does not believe the owner of it could get it to work in an emergency, if he wished. Not a single apothecary, as far as he knows, keeps one; and the writer does not know, among all his intelligent acquaintances, a single non-professional person who could use a stomach pump with success if he had a dozen of them at his command. A handful of salt and a tumbler of water can always be had; and anybody can mix a heaping teaspoonful of ground mustard with a cup of water, and get a person to swallow it. Either, swallowed, will empty the stomach; a "stomach pump" will do no more.

As has probably been observed, the *simplest* things, and those most likely of all others to be had everywhere, are the ones only spoken of in this pamphlet. The same purpose carried out at this point, leads us to say nothing now about the stomach pump; and, for the same reason, it possibly will not be referred to again.

Emetics.

For the purpose of rapidly emptying the stomach, in the decided majority of cases, before the arrival of a physician, and after it too, there is nothing like an Emetic. The easiest had, also, is usually the best. There are few places where these things cannot be had—*Ground Mustard, Common Salt,* and *Warm Water.*

Ground Mustard.—Take a tablespoonful, mix with a tumbler of water to about the consistence of milk. Give the person one-fourth of it at once. Then follow with a cup of warm water. In about a minute, give the person the same quantity again, followed by the warm water. If vomiting does not take place, continue giving until it does, letting a minute or so pass between each dose. Plentiful drafts of tepid water materially *assist* the action of the emetic, and the free use of it should, therefore, not be omitted. Mustard is not only useful as an emetic, easily found, and as readily given as anything else, but it is *stimulating* in character. This feature gives it a peculiar value in most cases where an emetic is demanded, for there is often, with the necessity for its use, a stimulant needed. The amount derived from Mustard is not always enough; sometimes it is; but when not, so much has been contributed.

Common Salt is even easier had than ground mustard, and is as certain in action. It is given, a teacup of water with as much salt dissolved as the water will hold, every minute or so, until vomiting occurs.

Warm Water, given cup after cup, is a safe emetic; but as the above mentioned articles are so easily had, it is rarely relied on alone for the purpose. Usually it is given to assist the action of the others, on the principle, perhaps, that a distended stomach is often easier emptied than an empty one. After vomiting has occurred, frequent drafts of warm water are often given to cleanse out the stomach. In many instances, for this purpose, warm milk, gum arabic water, flour and water, the white of an egg in a teacup of tepid water, and such substances, are given instead, with the expectation that their gummy, viscid properties fit them to entangle and detach particles of the poison adherent to the mucous membrane (lining) of the stomach. Besides, they are soothing to the perhaps irritated condition of the parts.

Tickling the inside of the throat by the finger, or with the tip of a feather, in many instances alone, will induce vomiting. Usually, after an emetic has been given, this is used to hasten its action.

S u l p h a t e o f Z i n c is another valuable emetic, often found in private houses. As much as will lie heaped up on a common two-cent piece is twenty grains, which is a dose, when dissolved in water. This quantity should be given at a single draught, followed by a cup of tepid water, and repeated every three minutes until three or four doses have been taken, or vomiting occurs. If there is none in the house, send to the nearest apothecary for sixty grains of the Sulphate of Zinc ("White Vitriol"). Empty the package containing this quantity into half a pint of tepid water. Stir rapidly with a stick, and it will soon dissolve. One-third of this half-pint should contain, of course, twenty grains of the sixty put in, and that quantity should be given at a single draught, followed, as all emetics should be, by draughts of tepid water. In a few minutes repeat, as directed about Mustard, unless profuse vomiting takes place.

P u l v e r i z e d I p e c a c u a n h a is another valuable emetic, particularly for children. It can be had of any apothecary by a messenger. Sixty grains (one drachm) of it may be requested. It is a ground root, and, as would be expected, does not *dissolve* in water, but mixes with it, like ground Mustard. One-third of the sixty grains, which is twenty grains (as much in bulk as will heap up a two-cent piece) may be given, mixed with a small teacup of tepid water, followed by a draught of tepid water. In a few moments, if vomiting does not occur, give another third, as you gave the first, to be followed in sixty seconds more by the last.

A good deal of trouble is often experienced in getting the person to *swallow*. This may be due to insensibility, fright, or stubbornness. The thumb of each hand may be slipped in outside and close against the teeth, along the line of junction, until the spot is reached behind where there are no teeth. Then through that vacant space slip the tips of the thumbs in between the jaws, when a separation can be readily effected. The thumbs should be kept there, for the patient cannot bite the attendant while his fingers are in such a position, and the handle of a strong iron or silver spoon, or piece of smooth stick, thrust back far enough to forcibly depress the tongue. The liquid can then be poured down the throat, if the person is lying on his back. At first it will fill up the space at the base of the tongue, but a little more depression of the tongue by the leverage given by the spoon or stick will cause it to run down the throat. There need be no fear of the fluid getting into the windpipe, for a very sensitive valve over the entrance of the trachea (windpipe) amply protects it.

The *first* vomiting, as said before, does not necessarily *clear* the stomach of its contents. Much of the poison may remain *adherent* to the mucous membrane, requiring frequent *washings*, as it were, for detachment and removal. After the first vomiting, there is usually little trouble in keeping it up, by simply giving plenty of tepid water. Warm water alone is often, as said above, an Emetic; and when none of the mentioned things can be had, must be wholly relied upon for the purpose.

Before the action of an Emetic can begin, a portion of the poison usually escapes from the stomach into the contiguous bowel. No vomiting can affect it; so, after the contents of the stomach have been removed by the action of the Emetic, it is always well, if the poison belongs to what is called, for convenience, the Mineral class (pp. 83), to administer good quantities of milk, which, passing down engages the activity of the poison. Flour and water will answer, but what is better, perhaps, is the white of eggs, mixed with water.

Now we will suppose all the *poison* has been *removed* by the above efforts from the stomach. The next thing is the removal of the *consequences* of the portion of the poison which has already commenced its work. If the mucous membrane has been injured, it should have rest from its usual work—digesting food—and be treated by suitable soothing applications, as barley water, gum arabic water, and such things. This should follow, where the poisoning is due to any of the articles embraced in the first class of substances treated of.

POISONING BY MUSHROOMS.

Persons not well acquainted with the differences between the poisonous and edible Mushrooms had better buy them of those who are, or go without. There are distinctions between them, but they are not of such a character as can be made evident in a place like this.

When poisoning from eating Mushrooms does take place, the contents of the stomach should at once be evacuated by an Emetic. (See page 77.) After vomiting has commenced, it should be promoted by draughts of warm water, barley water, but particularly by drinking copiously of warm milk and water, to which sugar has been added.

What has passed along into the bowels should be hurried out as fast as possible, by some cathartic, before further absorption into the blood can take place. Castor Oil might be peculiarly useful in such a case.

If there is much prostration of the strength, some easily-procured stimulant might be useful, as the Aromatic Spirit of Ammonia, or Brandy.

POISONOUS MEATS.

Eating meats of diseased animals is often followed by symptoms of a poisonous character. Animals in otherwise

6

perfect health, but which have been butchered and prepared for food after long and exhaustive confinement, are unfit for eating. Not only is the meat of such animals lacking in *nutritive* character, when compared with the meat of animals killed from the pasture without excitement, or after being kept until proper recovery from the effects of the journey to market, but it is much less savory, and shows a disposition to much more readily decompose. It might be here stated that it has been estimated by competent authorities, that between the two kinds of meat there is, in a commercial sense alone, as far as nutriment is concerned, a difference of nearly fifty per cent. in favor of the meat of healthy animals, butchered after complete recovery from the excitement and fatigue of drive or carriage to market. The additional cost per pound of meat to cover the expenses of extra care and precaution before butchering, would amount to but a small fraction of the percentage named, leaving the rest of it a true profit to the consumer.

The eating of this overdriven meat, it is said, is often followed by symptoms of irritation of the stomach and bowels; but they can, in the ordinary sense of the word, scarcely be said to be of a poisonous character, however much the use of them may derange the health.

POISONOUS FISH.

Several varieties of Fish, at *all* seasons of the year, are reputed to be poisonous. Of course, they should always be let alone. Should they have been eaten by accident, the best treatment is that given under the head of Poisoning by Mushrooms, p. 81.

Shell fish, at certain seasons of the year, after spawning, are considered poisonous when eaten; at least, they are unhealthy. This process of nature is known to be very exhausting to the individual, which during, or just afterwards, is so reduced in vitality as to be unable to resist ordinary tendency to decomposition.

Oysters, in hot weather, are often unwholesome, perhaps from the causes suggested; or it may be that the collection of liquid secretion between the shell and the contained animal, in hot weather, is in a state favorable to putrefaction upon slight exposure to the air; and the disagreeable symptoms often said to arise after partaking of this fish as food, is due to this as much as anything else.

MINERAL POISONS.

ALKALIES, ALKALINE EARTHS, ACIDS, METALS, Etc.

Ammonia.

The ordinary Aqua Ammoniæ, sometimes known as "Hartshorn," acts on the mucous membrane of the stomach, as would naturally be expected after knowing its effect upon the mucous membrane of the nostrils. When swallowed, it acts as a rapid corrosive poison. Owing to its pungency, it can scarcely be given by mistake in a state of purity. With Olive Oil, it forms the common "Hartshorn Liniment," and has thus been given internally.

A violently-acting corrosive substance, like Ammonia, leaves no time for Emetics. It is an Alkali, and the common dilute Acid known as Vinegar will neutralize it. Lemon juice also would answer the purpose.

Other concentrated alkalies, as Lime, Soda, and Potash, act in the same manner as Ammonia, and when taken internally must be combated in the same way, and with the same difficulties in view.

Mineral Acids.

The common Acids, Acetic, Muriatic, Nitric ("Aqua Fortis"), Sulphuric ("Oil of Vitriol"), are all *highly corro-*

sive in their action, unless largely diluted, and act with even greater rapidity when taken internally than when applied externally (see page 33). They are about as troublesome in this respect as concentrated Alkalies.

When taken, the Acid should be *neutralized*, as far as possible, by giving some harmless Alkali. Lime water is usually about as convenient as anything else for the purpose. Common Soap, from the Alkali it contains, might be given.

Antimony.

This metal is rarely accessible in its purity. One of its salts, as Tartar Emetic, or the Wine of Antimony (which is Tartar Emetic dissolved in Wine), is the usual source of the poison. Vomiting is one of the most distressing and prominent symptoms of poisoning by this substance. Assisted with copious draughts of tepid water, sugar and water, flaxseed water, much of the poison in the stomach may be gotten out. Another symptom is great prostration. If a small quantity only is known to have been swallowed, a teaspoonful of Paregoric in a little sweetened water may be divided into three portions, one of them being given every ten or fifteen minutes. It soothes the irritated and excited stomach.

The A n t i d o t e usually recommended is Nut Galls, or Oak Bark in powder. Half a dozen of the former, finely powdered, may be given, mixed with water. The active principle in each of these is what is called *Tannin*, now to be had of every Apothecary and Dyer. Ten grains of it (a teaspoonful—it is very light) in water will be equivalent to the Nut Galls mentioned. A strong infusion of common *Green Tea* contains enough tannin to make it useful as an antidote. An insoluble, and perhaps inert, Tannate of Antimony is formed.

Arsenic.

In some places this is called "Ratsbane," and poisoning often occurs from it. The Yellow Sulphuret of Arsenic (Orpiment), the Red Sulphuret of Arsenic (Realgar), and the Arsenite of Copper (Paris Green), are used as paints, and have all been used internally with fatal effects. The latter, Paris Green, has lately been much used by farmers for destroying injurious insects among plants. All these sources of poisoning by Arsenic should be surrounded by every possible precaution to prevent them from being accidentally used.

Many "fly poisons" contain it, and what is used in medicine under the name of Fowler's Solution is a solution of Arsenic. Arsenic acts as an irritant to the stomach and bowels, in many respects like Antimony and its preparations. As soon as it becomes known that Arsenic or any of its preparations has been swallowed in poisonous doses, the poison taken should be dislodged from the stomach, as far as possible, by vomiting (see *Emetics*, page 77), assisted by the finger to the throat, or the feather part of a quill. Free drinking of Milk, White of Egg and Water, or Flour and Water, should be encouraged. Not only do these things encourage vomiting, and tend to dilute the poison, but they offer something upon which the poison can expend its energy, to that extent saving the living tissue, and at the same time they tend to envelope the particles of the poison until the mass can be removed from the stomach.

The A n t i d o t e to Arsenic is the freshly-prepared *Hydrated Peroxide of Iron.* This can be had of any Apothecary in a few moments of time. It is quite harmless in character, and may be given in almost any quantity. The iron, in this particular form, combines with the Arsenic, forming a temporarily harmless preparation. This newly-formed compound should not be permitted to remain and be digested, but must be dislodged afterwards by an Emetic, which the bulk of the antidote favors.

Another Oxide of Iron, closely allied to, and said to be a useful substitute for, the Hydrated Peroxide, can be made by almost any one in a few moments, if some Aqua Ammoniæ ("hartshorn"), p. 00, and some of the common Muriated Tincture of Iron can be had. Both of these articles can be found in many houses, and if not there, in every Apothecary shop, or the office of every country practitioner.

Take a glass tumbler, or a graduated measure, pour in three or four tablespoonfuls (quantity not of much importance) of Aqua Ammoniæ, and then a tablespoonful or more of the Muriated Tincture of Iron.

A thick, dark, reddish precipitate, like brick-dust, is at once seen in the mixed liquids, which may be increased in quantity by gently stirring with a broom-splint.

This precipitate is the Oxide sought, and must be separated from the liquid by spreading a fine handkerchief or closely woven piece of muslin over a cup and pouring on the mixture. The liquid will run through, leaving the desired Oxide of Iron as a reddish-brown jelly-like powder. To free it from any excess of either substance used in its formation, a half pint or so of tepid water should be poured on in a gentle stream, to wash the precipitate. The washed precipitate is now ready for use.

A teaspoonful of this powder may be given every few minutes.

Calcined Magnesia and Pulverized Charcoal have also been recommended as antidotes in poisoning by Arsenic, but of their value nothing can be said by the writer.

Baryta.

This substance, largely used to adulterate certain paints, is sometimes accidentally swallowed in poisonous doses.

The Antidote is Water, acidulated to about the strength of lemonade with *Sulphuric Acid*, which converts the Baryta into an insoluble compound, which must be dislodged from the stomach by an Emetic.

Bismuth.

Some of the preparations known as "Toilet Powder" are largely made up of this substance, and when swallowed are followed by poisonous symptoms. The first object in such a case is to secure prompt and thorough evacuation of the stomach by an *Emetic*, followed by copious draughts of Milk.

Copper.

The most common form of poison from this cause is through the careless use of utensils made from it. Most Acids form soluble salts with Copper; hence Acids should never be used for cooking purposes in copper vessels. Many of the ordinary vegetables and fruits contain enough to form poisonous salts with this metal. Even Sugar, from the ease with which solutions of it are changed into Acids, should be cautiously used in contact with copper. When Copper is mentioned, it must be understood to apply to Brass, and other alloys into which Copper enters as a necessary component. Indeed, there is scarcely an excuse for the use of Copper or Brass as vessels any longer, owing to the superior advantages of similar vessels of Iron lined with porcelain, popularly known as " Enameled."

The stomach must at once be emptied by an *Emetic,* and copious draughts of Milk, or the White of Eggs mixed with Water. *Carbonate of Soda* (the ordinary Baking Soda of the kitchen may answer) is said to be an Antidote. As much as will lie heaped up on an ordinary nickel cent can be given every five minutes, in water, or in the other named liquids. *Iron Filings*, or the Ferro-cyanide of Potassium (*Prussian Blue*), in teaspoonful doses every three minutes, may be given.

Gold.

All preparations of Gold taken into the stomach act as the irritant poisons just described. The only difference in the treatment compared with that from poisoning by Copper is the Antidote. This is the *Sulphate of Iron*

(Copperas, Green Vitriol). As much as will lie heaped up on a cent may be dissolved in water, and one-third of the solution given every three or four minutes.

Iodine.

The common Tincture of Iodine, used for external application, is the usual form of this poison. There appears no Antidote, in the usual acceptance of the term. *Starch*, in water, may be freely given until vomiting is secured by an *Emetic*.

Iron.

The form usually taken is a solution of the Sulphate of Iron (Copperas, Green Vitriol). Its action is, like most of the poisons heretofore described, an irritant poison to the mucous membrane (lining) of the stomach and bowels. The A n t i d o t e is Carbonate of Soda. (See remarks upon this, under Copper, page 99.)

Lead.

The form from which poisoning by this substance usually takes place is the Acetate of Lead (Sugar of Lead). The Carbonate of Lead, the "White Lead" of the painters, and the Red Oxide ("Red Lead"), are also sometimes swallowed in poisonous doses. They all act as *irritant poisons*.

The treatment of such cases consists in giving, as an A n t i d o t e, water acidulated to about the strength of Lemonade with *Sulphuric Acid* ("Oil of Vitriol").

Sulphate of Magnesia (Epsom Salts), or the *Sulphate of Soda* (Glaubers Salt), in water, are also reputed antidotes. After the antidote has been given in poisoning by Lead an *Emetic* should be given.

When Lead is taken for some time, in any of its soluble

forms, in small doses, as when water has been kept in leaden vessels, or food kept or cooked in vessels "glazed" with lead, or the use of wines " sweetened" with the same metal; a peculiar train of symptoms slowly follows, known as "lead poisoning," or "painters' colic." All such possible sources of the introduction of Lead into the system should be carefully avoided; and as soon as the effects of the absorption begin to be suspected, no time should be lost in consulting a physician.

Lime,

If accidentally administered, acts like Ammonia (p. 99).

Mercury.

The Bichloride of Mercury (Corrosive Sublimate), often used as a solution in houses for destroying vermin about beds, is one of the most active poisons, when taken internally. The Red Oxide of Mercury (Red Precipitate) is another dangerous salt of the same metal. When swallowed, the *White of Eggs* should at once be given, and often repeated. In the absence of this form of albumen, common Milk can be used, or wheat Flour beaten up with Water.

These Salts of Mercury not only irritate the stomach, but so rapidly inflame and destroy it that some writers discourage the use of Emetics. If one can be given, however, before the poison has had time to produce these extreme results, there can be no objection to its use. The continued administration of the mentioned A n t i d o t e s is soon followed, as a rule, by free vomiting.

There really appears to be no excuse for keeping such Poisons about the house as Corrosive Sublimate or Red Precipitate. They are merely poor substitutes for good housekeeping.

Potash.

The Caustic Potash, in the form of common Lye, or the Concentrated Lye, when swallowed, acts as other Alkalies of the same general character. (See Ammonia and Lime.)

Nitrate of Potash (Saltpeter), in large doses, say half an ounce or more, taken internally, is followed by poisonous symptoms. There is pain, with heat in the stomach, vomiting, and purging of blood, great prostration, and other symptoms denoting the action of an *irritant poison*.

No antidote is known. The treatment consists in rapidly evacuating the contents of the stomach by an *Emetic*, and the free administration of *mucilaginous drinks*, with some Paregoric every little while, to allay the pain and irritation of the inflamed parts.

Phosphorus.

This is not often taken in a state of purity, perhaps. It is the active ingredient of most of the popular " Exterminators" for rats and other vermin. Children have been known to eat it with fatal results. They have also eaten the ends of common matches with similar consequences. Phosphorus acts as an irritant poison, inflaming the mucous membrane with which it comes in contact.

There is no antidote known. Some *Calcined Magnesia* may be given, in plenty of water, to be rapidly followed by an emetic, and then an abundance of mucilaginous drinks.

Tin.

Several compounds of this substance are used by Dyers, and have been used as poisons. They all act as irritant poisons. The treatment consists in copious draughts of Milk, White of Eggs in Water, and Flour and Water. Practically, the treatment advised under the head of " Copper" may be followed.

Zinc.

The Sulphate of Zinc (White Vitriol) may be termed poisonous in very large doses, were it not for the fact, constantly turned to good use, that it at once causes vomiting, and is brought up before damage can be done. Hence it is regarded as one of our most valuable Emetics.

Soda.

The same things are to be said about this Alkali as about Potash, Ammonia, and Lime. The rapid action of these substances upon the parts with which they come in contact leaves little to be done with much confidence.

Silver.

The chief source of this poison is the Nitrate of Silver (Lunar Caustic), either solid or in solution. Its action as a " Caustic" is well known, and it is in this manner that it acts upon the throat, stomach, etc., when taken internally, in solid stick or in solution.

Nitrate of Silver (Lunar Caustic) is the base of the numerous popular " hair dyes," and under this form has been accidentally and criminally taken.

The A n t i d o t e for the Salts of Silver is *Common Salt*, which immediately decomposes and destroys its activity. The rapidity and completeness with which this is done is seen in the well-known domestic precaution of preventing solutions of Silver employed as Indelible Ink from staining, by immediately, while the spot is moist, touching it with salt and water.

Alcohol.

Alcohol itself, or in the form of Brandy, Gin, Rum, Whiskey, taken in large quantities, is followed by symptoms of a violent poisonous character, and if relief is not at once had death often ensues. It is not an unfrequent occurrence for the press to report instances where children have swallowed Alcohol left accessible to their reach, and have died in consequence.

When quantities have been taken sufficiently large to be followed by alarming symptoms, the contents of the stomach should be evacuated without delay, by tickling the throat with a feather or the tip of the finger, by an Emetic, such as Ground Mustard and Water, Pulverized Ipecacuanha, or Sulphate of Zinc; or the stomach pump may be used, if convenient. The vomiting should be assisted by copious draughts of warm water, or other things of the kind.

The Alcohol in the stomach having thus been disposed of, the portion which passed from the stomach into the blood, and was carried to the Brain and the rest of the Nervous System, where its poisonous action is being exerted, should next claim the attention, if the symptoms appear to demand it. The action of Alcohol and its preparations upon the Brain and Nervous System is seen under the common name of Intoxication. This may exist as mere stupor, or the Brain and the Nervous System may be so completely overcome by the presence of such a quantity of the poison in the blood that the action of these parts upon the muscular movements of the Chest and the Heart is no longer kept up, and death ensues from Asphyxia, as described under that head.

VEGETABLE POISONS.

Most of the class of Poisons termed Vegetable act as Narcotics or as Acro-Narcotics. With some modifications, which will be mentioned in place, the treatment of *all* cases of *Narcotic* poisoning is essentially the same; and a similar statement may be made in reference to the treatment of all cases of *Acro-Narcotic Poisoning.* Hence, in speaking of the Vegetable Poison standing first in the alphabetical arrangement of these substances, the directions have been given under two heads; the nature of the Acrid or *irritating* symptoms and the treatment of the *Narcotic* symptoms. In speaking of the other poisons, in turn, to save space and avoid repetition, the reader will be directed, for details of treatment, to one of the substances, when directions are given in full.

Aconite.

Aconite is known under the names of "*Monkshood*" and "*Wolfsbane.*" When swallowed in an overdose, it is rapidly followed by symptoms known as acro-narcotic; in other words, irritating to the part, and narcotic to the Brain and Nervous System.

The T r e a t m e n t naturally would consist in getting out of the stomach all the poison there not already absorbed into the blood. In this Acro-Narcotic Poison we have two reasons in view for such a course; first, to avoid, as much as possible, the *irritating* features of the poison to the mucous membrane (skin lining the stomach and approaches), and, secondly, to prevent further *absorption* into the blood and narcotization of the Brain and Nervous System.

The contents of the stomach are removed by tickling the throat and base of the tongue by the finger or a feather. An *Emetic* (page 77) of Mustard and Water, Pulverized Ipecacuanha, or Sulphate of Zinc, Flaxseed Tea, Gum Arabic Water, Sugar and Water, Milk, White of Egg, or things of this general character, should be freely given at the same time, to *protect* the mucous membrane of the stomach from the irritating feature of the poison.

There is nothing known to neutralize or destroy the

poison in the blood thus acting through the Brain and
Nervous System upon these important organs of breathing
and circulation, so the efforts for relief must consist in
keeping up the respiration by artificial breathing (page
10) until the kidneys, skin, and other organs have had
time to eliminate (throw out) the aconite, and so little
remains in the blood as to no longer seriously interfere
with breathing and circulation. We shall, therefore, refer
the reader to " Asphyxia from Carbonic Acid Gas" (page
14), which is said to act in the same manner upon the
human body.   Also see " Opium" (page 97).

Belladonna.

Belladonna, or " *Deadly Night Shade,*" has been introduced into our gardens as an ornamental flower, and poisoning sometimes occurs from eating the berries or leaves.
Solutions of the active principles of this plant are used
under various forms as an application to the eye or brow,
for certain purposes, by the surgeon, and should never be
left where they can be swallowed by mistake.

Belladonna acts as a Narcotic poison, like Opium in
many respects, only there is dilatation, to a marked degree
of the pupil of the eye, and a peculiar redness or suffusion
of the face, which are not seen in poisoning by Opium.
This duskiness of the face is the symptom first observed
by the physician.   To discriminate between the two, it
may be remarked, that Stramonium (Thorn Apple, Jimson
Weed) gives results closely resembling Belladonna.   Both
of these substances, Belladonna and Stramonium, are attended, when swallowed in large doses, with a peculiar
dryness of the throat and mouth, delirium, not accom-
panied at first with *stupor*, like Opium, but with violent
gestures, often violent laughter, and a peculiar disposition
to pick in the air, or at the clothing, for imaginary objects.

Treatment.—The poison in the stomach must be immediately gotten out by an Emetic (page 77), and the treatment under the head of "Narcotics," in the preceding
page, followed out as seems required.

Bryony.

The root of this plant, when swallowed in sufficient
quantity, acts as an acrid, highly irritating poison. It is
quite a common plant in Europe, but less seen in the
United States.

If taken in poisonous quantities, *empty the stomach*
as soon as possible ; and, as in the case of *all* highly *irritating* poisons, this should be followed by free drinking of
Milk, Flaxseed Tea, White of Egg and Water, Sugar and
Water, Gum Arabic Water, and similar things.

Camphor.

When taken in large doses, Camphor acts as a narcotic
poison. The contents of the stomach, in such cases, should
be evacuated by an Emetic (page 77), followed by draughts
of warm water, Flaxseed Tea, Gum Arabic Water, Milk,
and similar substances. The strong odor of Camphor in
the breath and perspiration, in case of poisoning, with
narcotic symptoms, would naturally point out the peculiar
agent employed.

Draughts of strong *Coffee*, it would seem, might be as
useful in the case of narcotic poisoning from Camphor as in
the case of narcotic poisoning by Opium and other substances of this class.   (See "Opium," page 97.)

Conium ("Hemlock").

This well-known plant is an active poison, when taken
internally in sufficient quantity. It is supposed to be the
narcotic used by the Greeks, and other nations, to destroy

the life of condemned criminals. Socrates and Phocion, it will be remembered, were compelled by the Athenians to drink a decoction of it.

Its action is that of a *narcotic*, and the treatment consists in getting out of the stomach, as soon as possible, by an *Emetic* (page 77), all the vegetable there, and then counteracting the effects of that which has been *absorbed* into the blood, and tends to overpower the brain and nervous system.

See remarks on "Alcohol" (page 91), and "Opium" (page 97).

Digitalis ("Foxglove").

This beautiful plant of the garden, cultivated in this country for its flower, and used, in proper quantities, as a valuable medicine, is a poison of the *narcotic* class, with a disposition to overcome the portion of the nervous system controlling the action of the *heart*.

The same *treatment* should be pursued, when Digitalis has been taken in poisonous quantities, as recommended for other narcotics. The peculiar tendency to stop the action of the heart should be specially combated by giving twenty or thirty drops of Aromatic Spirits of Ammonia every three or four minutes in a tablespoonful of water, or some other stimulant.

Dulcamara ("Bitter Sweet," "Woody Nightshade").

This well-known plant belongs to the narcotic class of poisons, with symptoms like those of Belladonna and Stramonium.

The treatment is about the same as heretofore spoken of since we began our remarks upon the Vegetable Poisons. See "Belladonna," page 94.

Hyoscyamus ("Henbane").

This vegetable, made use of in medicine, if taken internally in improper quantities, acts as a narcotic poison, like others of the same natural order (Solanacæ), as Belladonna, Dulcamara, and Stramonium.

As the treatment of a poison depends upon the action of the agent, we would naturally treat poisoning by Hyoscyamus as by other articles acting the same way. See "Belladonna," page 94; "Stramonium," page 104.

Lobelia ("Indian Tobacco").

This vegetable is not now much used as a medicine by physicians, as the comparatively recent discoveries in chemistry have added substitutes to the list of drugs, without the peculiar disadvantages of this substance.

In poisonous quantities, Lobelia belongs to the class of acro-narcotics spoken of under the head of "Aconite" (page 93). Fortunately, one of the symptoms following its use is vomiting. This should be encouraged by drinks of tepid water, Gum Arabic Water, etc.; and, if kept up until all the poison is rejected by the stomach, a favorable issue may be expected. Should vomiting not occur at *once*, as a symptom, enough of the poison may be absorbed into the blood to exert a fatal *narcotic* influence upon the brain and nervous system; or, perhaps, to speak more precisely, through these organs upon the movements of respiration and circulation of the blood.

Opium.

This substance, or the numerous preparations of it used in medicine, is one of the most frequent causes of poison a physician is called in to see. A fruitful source of mistake is that of confounding together of Laudanum and Pare-

7

goric. When the latter is supposed to have been given by
the nurse, the mistake is not found out until it is often too
late to be of material service in averting a fatal end.
Morphia, the active principle of Opium, is often kept as a
solution, in private houses, for domestic use.

Any of these preparations of Opium, in improper doses,
are followed by symptoms of *narcotic* poisoning. Not
only these, but many popular nostrums, as "Infant Cor-
dials' and "Soothing Syrups" of various kinds, depend
for their utility upon some preparation of Opium, and are
hence often followed by symptoms of narcotic poisoning.
None of these things should be used. If a child cries, it
is usually because it feels *pain;* and, instead of making it
so stupid with narcotics that it cannot *feel* the pain, it is
better to go to work and find out the *cause* of the pain,
and remove it. There is some reason for the suspicion
that, in many instances, where a modicum of the popular
remedies of this class are not furnished by the mother to
the nursery, the enterprise of the nurse, "rather than see
the child suffer," secures it for the charge committed to
her care, from a neighboring apothecary shop.

Opium, its preparations, and the active principle of the
drug, Morphia, all act in the same way; by absorption
into the blood, and distribution by it to the brain and
nervous system. Through these organs the movements
of the chest and heart become more or less interfered
with. In this respect its action is essentially like that of
Carbonic Acid Gas, Alcohol, and most of the vegetable
poisons herein described, without, however, any acrid or
irritating complication.

T r e a t m e n t.—It is safe to say that at present there
is no known Antidote to any of the narcotic poisons, using
the word Antidote as understood by physicians. What is
in the stomach must be taken out, to prevent further ab-

sorption, and what is in the blood must be worked out, under proper guidance, by the processes of nature constantly engaged with such products. If the breathing and circulation tend to cease, because of the inability of the brain and nervous system to temporarily discharge these duties, these essential movements must be taken charge of by a friend.

An active *Emetic*, like Ground Mustard, must be given at once, remembering that trouble may be found in getting it to *act*, owing to the diminished sensibility to its presence, from the local stupefying action of the Opium to the mucous membrane of the stomach. The action of the Mustard should be assisted by tickling the inside of the throat with the fingers or a feather.

Sulphate of Zinc, Salt and Water, Pulverized Ipecacuanha (page 79), may be given; in fact anything, t o e m p t y t h e s t o m a c h a s s o o n a s p o s s i b l e.

The narcotic effects upon the brain, at the same time, as far as possible, must be attended to. If the *respiration* is yielding to the poison, that is, falling much below the standard of about twenty to the minute, it must be sustained by assistance. As directed under the head of "Asphyxia from Carbonic Acid Gas" (page 14), the exposed body of the patient should be *dashed* with cold water, not neglecting the head, face, and chest. After the cold water has been sufficiently used in this way, the body should be dried, removed to a dry spot, and hot applications made to the extremities and other parts. This is necessary, owing to the heat-producing power of the body being impaired by the suspended or diminished respiration.

If the respiration is not *suspended*, but is going on at a *diminished* rate, say six or eight to the minute, artificial respiration is not required, unless the number of respiratory movements of the chest falls below that; but the other measures may be used. In addition to these, a strong *stimulant*, in the shape of twenty or thirty drops of Aromatic Spirits of Ammonia in a tablespoonful of water, may be given three or four times, at intervals of a couple or more minutes. It is better than brandy, or anything alcoholic, because the mode of the action of brandy is much the same upon the brain as Opium, and it might be rather adding to instead of taking from the poison that is at work. The often referred to Aromatic Spirits of Ammonia will give the advantage, without the suggested disadvantage. A few tablespoonfuls of *very* strong, freshly-made *Coffee* is a useful thing to give in such cases.

Among measures to *keep in activity* the circulation and respiration, as well as to promote the elimination (casting out) from the blood of the poison acting as a narcotic, there are few things more useful than *muscular exercise*. If the circumstances permit it, this is often effected by a person getting on each side of the individual under the influence of the narcotic, supporting him under the arms, and walking up and down the floor with him. The writer saw a case where the person under the influence of this narcotic (Opium) was wholly unconscious, and with breathing not over six to the minute. A relay of persons walked him up and down a long room for three hours, a person walking behind to hold the head of the patient in a natural position over his shoulder. Occasionally he was stopped at a suitable place, the blanket around him removed for a moment, and he was dashed with cold water. The body was then rapidly dried, and the blanket in the meanwhile having been heated, was reapplied. · Once the respiration became so feeble that he was placed on his back, and the artificial breath (Silvester's method, page 10) used for some minutes. While this was being done by one person (a policeman in this case), under the directions of the physician, another individual caught hold of each ankle, bent the knee, pushed the knee upward until it touched the stomach, and then straightened out the entire leg. This was done several times a minute, and perhaps was as useful as walking the patient up and down the floor, besides not interfering in any way with other measures of relief. Indeed, there is some reason for thinking the action of the heart would be more favored by such muscular movements with the patient on his back, than if the person was standing upright.

Whipping the body by a folded towel wrung out in cold water is of the greatest use in cases of narcotic poisoning.

In case medical assistance shall not have been secured, and the patient shows signs of improvement, in the shape of more frequent respirations, stronger pulse, and returning consciousness, many of these measures may be omitted as the apparent necessity disappears. In a short time the patient will appear as a person who is soundly sleeping from the effects of a full dose of Opium or other narcotic; the quantity *beyond* that having been parted with by the blood. He may now be let alone, unless some return to the previous condition is noticed, when a dose or two of the strong and easily procured stimulant, Aromatic Spirits of Ammonia, may again be given him.

It must be recollected that a person who has been in such a state as to require all these artificial muscular movements is, practically, in about the condition, as far as strength is concerned, of a man who has run hurriedly several miles without resting. He, of course, has consumed all his available *strength*, and the sooner it is made up to him by beef tea, and such things, the sooner he will be where he was before the narcotic was taken.

Oxalic Acid.

This substance is largely used in the arts, and in private households, for removing stains of iron from textures and surfaces, which it does by combining with an otherwise insoluble salt of iron, and converting it into a soluble oxalate of iron, easily removable by water. From the strong resemblance Oxalic Acid bears to Epsom Salts, it has often been taken instead of the well-known purgative of that name. To avoid the possibility of such an accident, Oxalic Acid should be kept in another part of the house from where medicines are kept, and no precaution omitted, by label and other marking of the parcel, to make the difference between them as decided as possible. It is well to remember, also, that, wholly unlike Epsom Salts, the taste of Oxalic Acid, applied to the tip of the tongue, is quite *sour*.

When swallowed internally, the activity of this poison admits of no delay. It belongs to the class of irritant poisons spoken of so often, and produces death, it is said, by destructive action on the mucous membrane (lining) of the throat, stomach, and bowels.

Time can scarcely be lost to give an Emetic; but something must be given to rapidly combine with it, and divert its activity from the parts mentioned. It has a strong affinity for Lime, forming with it a comparatively insoluble Oxalate of Lime; and for Magnesia, forming with it an insoluble Oxalate of Magnesia, which can be dislodged with less haste. A teaspoonful of the Lime at the bottom of a bottle of Lime Water, when made as directed in another place (page 116), mixed with a cup of water, might be given every few minutes, or some crushed Chalk (a Carbonate of Lime), or some Magnesia, may be given. All these things can easily be had, and not a moment need be lost in getting the person to swallow them. The common " whiting," used for polishing glass, making cheap paint and putty, is essentially the same as prepared Chalk.

After the Oxalic Acid is supposed to be neutralized, an Emetic of Ground Mustard or Pulverized Ipecacuanha may be given.

Pulsatilla.

The eating of this plant, " Meadow Anemone," or parts of it, has been followed by symptoms of acro-narcotic poisoning. The plant is so active at times that when applied externally, irritation to the parts touched is felt. When poisoning results from swallowing it, the course of treatment recommended under " Aconite" (page 93) may be followed.

Sanguinaria (" Blood Root"),

Taken internally, in an overdose, acts as acro-narcotic poison. See " Aconite" (page 93).

Savine.

This is an active irritating poison, inflaming the stomach and bowels. When thus taken, vomiting, by tickling the throat with the finger or a feather, should be at once induced. The mucous membrane (lining) of the bowels should be protected from the irritating action of what has escaped beyond the stomach before it could be emptied by vomiting, by drinking large quantities of water or milk, with good quantities of Gum Arabic dissolved in it. If the *Oil* of Savine, which is the usual form of the substance when used with a criminal intent, has been taken, it might be well to take a dose of Castor Oil.

Spigelia.

The use of this plant, commonly called " Pink Root," as a destroyer of worms, was given, it is said, to the whites, by the Cherokee Indians, and has become very general throughout the entire country. It is given with a great deal of confidence and recklessness, and is often followed by symptoms of a narcotic character, attended also with convulsive movements. When such poisonous symptoms follow its use, *vomiting* should be promoted, and kept up by frequent draughts of warm water. As in the case of other narcotics, a drink of strong coffee may be of service. Acidulated drinks, as water and vinegar, water with lemon-juice, are thought to be useful, and probably are, in favoring the elimination (throwing out) of the poison absorbed into the blood, by the action of the skin and kidneys, which they promote.

Stramonium,

Usually known as "Thorn Apple," or "Jimson Weed," belongs to the same natural order in botany as Belladonna, Dulcamara, and Hyoscyamus; and when taken internally, in improper quantities, is followed by *similar* general symptoms. Children often gather the seeds and eat them. A history of the case, the evidence of some of the seeds or capsules, the narcotic symptoms, with the peculiar duskness of the face and dryness of the mouth and throat mentioned when speaking of Belladonna, are sufficient to point out the vegetable used. There is a decided disposition to laugh, and pick at imaginary objects, on the part of the person under its effects.

Treatment has been given under the head of "Belladonna" (page 93).

Strychnia.

This is the active principle of the Nux Vomica, or "Dog Button," as it sometimes called, from the use often made of it. The action of this poison is so rapid that, like

Prussic Acid, little can be done to delay death. This poison acts in a peculiar manner upon the nervous system, throwing the muscles of the body into *strong* convulsive movements. The convulsions from Strychnia are attended with one strongly-marked and peculiar feature. It is a disposition, during the convulsion, for the heels and the back of the head to meet (opisthotonos), under the influence of the violent muscular movements. Whenever *this* is seen, and if seen it will surely be remembered, the coincidence between it and the use of Strychnia should be remembered.

The stomach should be evacuated with the least possible delay, if it is known the person has just taken the poison. If convulsions have occurred, and death taken place, it may be well to remember that death resulted from Asphyxia, the spasmodic action of the muscles attached to the ribs having prevented movements of respiration. Artificial breathing, in such a case, should be *tried*, with the hope that something might possibly be done to invite back the natural movements.

Tobacco.

To a person not accustomed to its effects, by beginning with small quantities, and persisting in its use, tobacco is an acro-narcotic poison, agreeing in its essential characters with Aconite, and others of the same general class. The movements of the heart become so much interfered with that death may take place unless proper assistance is at once given. Fortunately, like Lobelia, it acts as an Emetic, and before enough can be absorbed into the blood from the stomach, the contents of that organ are rejected. Hence, when death has ensued from the direct use of Tobacco, we find that it was used as an injection, a form in which it should *never* be given.

Other Vegetable Poisons.

Besides those enumerated in the foregoing pages are many others, whose names even cannot be given here. Most of them belong to the *Acro-narcotic* class, and may be treated as advised in speaking of those mentioned under that head. See Aconite (page 93).

SIGNS OF REAL DEATH.

Usually it is not a difficult matter to pronounce with confidence whether a person is really dead, or whether it is an instance of what is called suspended animation; but sometimes it becomes a question not easily determined, even with professional assistance.

There are few fears, perhaps, as widely and as universally entertained as the fear of being buried alive; and there is probably no apprehension which, after a careful and extended examination, has as little in fact to support it. Where it has become necessary, in some few reported instances, to examine the remains after burial, the change in position of the body from that in which it was supposed to have been at the moment of interment, is doubtless due in most instances to the formation and sudden escape of gases, the result of decomposition, from cavities of the body. Any one who has attended a funeral and observed the movements necessarily given the casket in taking it from the house to the cemetery, must see how readily the cylindrical form given the corpse by the conventional manner of pinioning the arms to the chest, and the feet to each other, permits it to be influenced in carrying; and it is rather a matter of surprise that the expected position, flat on the back, is as often found as it is. Either of these explanations will account for the change of position sometimes seen after burial, without for an instant calling in the dreadful and unjustifiable supposition that burial had taken place before life was extinct.

In many countries in Europe, where the remains of persons deceased are exposed under official inspection for some time after reported death, statistics of the most reliable character, extending over an uninterrupted period of many years, do not reveal, among the hundreds of thousands thus placed, that one has ever afterward shown a sign of life.

While it is possible that a person might be supposed to be dead and yet not, the usual method practiced in cities by undertakers, carefully and completely surrounding the body with ice as soon as they can, is well calculated to do away with the possibility of being buried alive. It is not an unheard of thing for the remains to be so packed within three hours after supposed death; and that, too, without any reason being apparent for the indecent haste, except the possible convenience of the men sent by the undertaker.

Under scarcely any circumstance should this be permitted, especially in the case of those who have died on the highways, at hotels and other places away from relatives and personal friends. During the prevalence of fatal epidemics, particularly of cholera, where there is fear of contagion, every known precaution should be taken to prevent even the most remote possibility of the thought ever afterward arising that undue haste occurred in placing the remains beyond friendly assistance in case it could have been of use.

Among the tests usually applied to a person supposed to be dead, is the absence of sensibility. While it is true that sensibility to punctures, pinches, blisters and burns demonstrates that the person is not dead; the want of sensibility to such things only proves that the individual does not feel them ; or feeling, is incapable of responding to them.

Another test is the absence of circulation as revealed by the action of the heart to the hand of a bystander; or the absence of the impulse of the blood by that organ to the artery at the wrist, properly known as the pulse.   There is little real value in the absence of both of these signs. The heart may be acting feebly, or the attendant unable to detect the movement.

The cessation of breathing is also often relied upon. There may be no movement of the walls of the chest, as is often seen in persons just brought from the water, exposed to carbonic acid and other gases, who afterwards revive without even assistance.   Besides, the method employed for learning whether breathing takes place may be unreliable.   A mirror held for the purpose near the mouth, to collect the moisture of the breath, does not always reveal it even when there ; nor is the force of the breath always strong enough to deflect the flame of a candle held near, nor to give movement to a wisp of cotton held near the lips.

It is sometimes thought that the person is dead when he is cold, and alive if he preserves his warmth.   A moment's reflection will show how little reliability can be placed in these signs.   For instance, the drowned are often, and the frozen always cold ; and both have been restored.   It is said, too, that those suffocated by strangulation or inhalation of certain gases preserve their heat for some time after undoubted death, as long even, as twelve hours after unconsciousness has commenced.

The general appearance of the face, the softness, sinking and relaxation of the countenance, and dimness of the eyes have all been considered, from the days of Hippocrates, as furnishing valuable signs of death.   While of recognized value to the mind of the physician, they cannot, of themselves, be *wholly* relied on for the purpose, but must be regarded when properly interpreted, as only presumptive evidence of the possibility of near or complete death.

In this connection it ought to be remarked that the commonly resorted-to test of suspected death, of exposing the eye to the light, is of much less value in establishing the fact of complete dissolution than is commonly imagined. Life may be present, but the eye as lacking in sensibility to light as other organs to their proper stimuli; besides, contraction of the pupil may be already so complete that more should not be expected.

One of the most reliable signs of death evident to ordinary observers, is the peculiar stiffness or rigidity of the body (*rigor mortis*), but as a sign closely resembling it is sometimes seen in life, as well as in suspended animation, certain essential points of distinction between the two must be kept in mind to establish the difference between them.

If the contraction or rigidity of the muscles is due to their convulsive action, instead of being the stiffness of death alluded to, considerable difficulty will be found in changing the position of the limb, and after it has been done there will be a constant disposition to revert to its former state. In death, however, the limb is apt to remain as last placed.

In the peculiar nervous condition known as Catalepsy the tendency for the limb to remain as placed is likewise seen as it is in the *rigor mortis*, and in some cases of this disease professional advice may be necessary to decide the difference between them.

When the body has been subjected to the influence of cold, as when the person is "frozen," a stiffness like that of death is found, but it affects not only the muscles different from the stiffness of death, but extends in like degree

to the abdomen, breast and other organs. Besides, when
the position of a frozen member is changed, a slight crack-
ing noise is made and felt, caused by the movements against
each other of the atoms of ice contained in the part.

If from any cause the person supposed to be dead is
cold and soft, whilst a certain degree of stiffness ought to
be seen, interment should not take place, but might be
postponed until a physician can make a satisfactory examination of the case.

All authorities, however, agree upon the reliability of
one symptom of death; and it may safely be said, the only
reliable one. It is well-marked putrefaction.

By this is not meant the *appearance* of putrefaction,
but the undoubted fact of it. In some forms of dysen-
tery a peculiar cadaveric odor is present, but it does not
by any means imply that death has taken place, nor even
that it must. The same suggestive odor is seen when
gangrene of a limb has occurred, or destructive ulceration
is going on. Purple blotches of the skin, with other
signs of decomposition, are occasionally met with, occurring
with other supposed signs of death, without death really
having taken place, or in fact a likelihood of it.

The putrefaction under consideration cannot well be de-
tected by the unprofessional, but the question should always
be submitted to the skill of a medical man. It usually
makes its first appearance over the abdomen, close down
toward the groin and extending upwards. At these and
contiguous points the skin first turns a dusky yellow, soon
merging into a greenish tint, more or less mottled, and
in a short time becomes softened to the touch, with the
evident odor of decomposition. The color alone is not to
be depended on, but the mentioned later stages, of what the
color may be the beginning of, should positively decide the
case.

APPENDIX.

In a number of instances it has been found necessary to speak of a few articles of medicine usually kept on hand in private houses, factories, and places where accidents are likely to happen.* It is scarcely necessary to point out how great a risk is assumed in not keeping a little supply of such things on hand, ready for use at a moment's notice. Most prudent people do keep them, although physicians know of cases where an apothecary has been rung up of a cold night to supply ten cents' worth of paregoric for a sick child. There is scarcely any excuse for such effrontery. Often there is a disposition to purchase these domestic medicines wherever they are said to be kept, and at the lowest price. This should never be done with articles intended for internal use, as the education of the buyer does not permit him to judge of the purity, strength, or activity of the drug. They should invariably be bought of the most reliable apothecary or retail druggist, if in the city or large town ; or if in the country, at the office of a reputable practitioner of medicine, who has made, or will guarantee the named essential features of the different articles.

What is left unused of prescriptions ordered by the physician, should not be preserved, as there is not one chance in a hundred that the same special combination will ever be required again; unless it is some liniment, or constituent of one, which can be utilized. Besides, as a rule, medicines do not keep well ; and the more bottles of such things there are about the house, the greater the chance of a mistake in getting hold of one, when another is needed.

But a small quantity should be bought at once, for the reason just stated, that most of them undergo changes in character after being kept on hand some time. Each substance should be kept in a flint glass bottle with a closely fitting cork stopple ; or, what is better, a carefully fitted ground glass stopple, as many medicines erode the

---

* Wherever spoken of, reference is had to what are called the standard articles prepared according to the United States Pharmacopœia, a volume revised every ten years, containing the accurate standard for the regulation of the strength and purity of medicines used by the physicians of the United States, and everywhere recognized by them as an authority upon all matters therein contained. It bears as close a resemblance as practicable to similar works, of the same obligatory character, of Great Britain and the different countries of Europe. This is the reason why physicians write the names of medicines and directions for preparation in Latin, so that wherever presented, no doubt can arise, from comparison with the works, as to what the writer intended.

111.

delicate cork, permitting the escape of a valuable element or giving access of the air to the contents of the bottle. Each bottle should likewise be correctly and distinctly marked with a printed label; and when the medicine is poured out, pour from the side of the neck opposite the label, so the last drop, if any, will not trickle down upon the label and deface it.

When not in immediate use, all medicines should be kept in a separate closet or other well defined space, where nothing else is kept, unless the little appliances of domestic surgery. It should be wholly free from dampness, as moisture impairs or destroys the activity of most drugs, especially those in powder, if it can get access to them. If the closet can be kept under lock and key, so much the better. Light must be excluded, as it destroys many substances, as the sweet spirits of niter; and injuriously modifies the character of the oil of turpentine. A low uniform temperature is likewise needed, otherwise the heat will vaporize the alcohol or ether of many preparations, and the supposed strength of the article may be dangerously interfered with. Besides this, heat destroys many medicines.

Syrups, or medicines containing sugar, when poured from a bottle, care should be taken to keep the neck at the cork free from the mixture. Independently of the neatness, the stopple and neck of the bottle should be wiped on each occasion of use, to prevent the collection and decomposition of the saccharine matter, as the character of the medicine may be modified by it.

Liniments for external use, in the majority of cases, and the same is true of Lotions, depend for their usefulness upon articles not to be taken internally. They should be kept in a distinct corner by themselves, labeled "poison," and as soon as used, returned to the spot they belong. Many of them, containing chloroform, ether, or other volatile substances, are apt to dislodge the stopple, and this element, perhaps the active one, rapidly evaporates. When ordered for immediate use, these substances can scarcely be preserved in their proper proportions to the other ingredients until consumed, even when an attempt is made to tie down the stopple.

Powders should be kept in tight metal boxes ; or, what is better, wide mouth bottles with closely fitting ground glass stopples.

Ointments ("salves") should never be kept in any quantity, or for any length of time. Unless purchased in large cities, at the retail shops, where large and rapid sales compel a constant renewal of stock, they are usually rancid. This is especially true of what is called "cold cream." A rancid ointment is unfit for use to a delicate part. It is quite as difficult, as a general thing, for an apothecary to keep ointments from becoming rancid, as it is for a housekeeper to preserve butter in its original freshness.

Pills kept for some time, particularly those containing certain articles, become so hard that they are about as soluble in the stomach as grains of coffee. This is especially true of sugar-coated pills.

This defect can be overcome by inclosing the number to be taken in a piece of muslin and reducing them to fragments by a blow.*

Medicines to be taken internally are usually ordered by drops, teaspoonful, dessertspoonful, or tablespoonful; not because these measures of quantity are correct, but because they are convenient. Drops vary in size, according to the temperature of the liquid, the shape of the edge over which it is poured, and the specific gravity and general character of the substance poured. All other things being equal, drops of ether are, perhaps, not more than one-third as large as a drop of syrup. However, the physician usually takes all these things into consideration when medicines are ordered, so if the directions as to the number of drops is followed, no danger need be apprehended from this source.

A Teaspoonful means the quantity occupying the space occupied by forty-five drops of pure water. Some teaspoons now the style hold much more than this, hence in every house should be kept a teaspoon known, by exact measurement, to contain just a typical teaspoonful, or forty-five drops of water of ordinary purity. With this correct standard another can be found holding just twice as much, and this will be what is called, in measuring medicines, a **Dessertspoonful**. This doubled, or four teaspoonsful, is the **Tablespoonful** of medicine, or half an ounce.

What is best of all, is the common **graduated measure**, as it is called, used by apothecaries and all who wish to accurately estimate quantities of liquid. Not only the technical characters should be cut on the glass, but the quantities written out in full in plain English. There is a kind in the market with the characters pressed on the glass instead of being etched, but some think them, as a rule, less accurate than the other kind. If a true one cannot be had, take some water and count out forty-five drops until a "teaspoon" is found of the precise size to hold it.

* There are few observant physicians who will not say that these things are used much too often when not really necessary.

Many people take them because they are what they call "bilious," which, in nineteen cases out of twenty, and this is said thoughtfully, means that the person has eaten too much food; not diminished the consumption of heat-producing articles, as fatty matters, as the heated weather approached; or has taken food which did not digest properly, owing to its nature, the mode in which taken, or the condition of the stomach at the time. In such cases, absence from food until the material already there can have passed off, and the alimentary tract restored, is all that is needed. A little thought afterwards, with a little manly forbearance, is all that is needed to keep from getting "bilious" again.

Another thing is to "purify the blood." But how do you know that it needs it? Read over what has just been said about "bilious" people, and see if there is not some suggestion there to meet the case.

Another thing often alleged is "constipation." Now there is often apparent reason in this. Physicians will say, however, that what is a necessity for one person twice a day, will not be for another twice a week; and one enjoys as good health as the other. If the reader has reached the age of forty years, and will think back over things, he will find things which were once a necessity have become useless; and things once not needed, cannot now be dispensed with. Therefore, because things do not happen now as they once did, or happen now that once did not, remember there may be nothing wrong about it; at least, not enough to excuse the wrong of taking a lot of pills without consulting a physician as to whether they are really needed.

8

One teaspoonful = 1 drachm = forty-five drops pure water.
One dessertspoonful = 2 drachms or 2 teaspoonsful.
One tablespoonful = 4 drachms = 4 teaspoonsful or 2 dessert-spoonsful, and is also equal to one half of a fluid ounce. Two table-spoonsful, of course, make one fluid ounce.

While speaking of certain articles of domestic medicine which should always be kept on hand, especially should several large and reliable apothecary shops not be in the immediate neighborhood, there is one practice to be decidedly condemned.

For some reason or another, or perhaps more correctly to say for lack of reason, certain persons persist in keeping about the house a parcel of Arsenic, Corrosive Sublimate, and if particularly favored, a few grains of Strychnia. They are often purchased under the delusion that they are intended " for rats," or something else. Some-times, even, they are carried around in the waistcoat pocket, or kept on the mantel, or in a well-known conspicuous place on the clock, to be displayed and their merits dwelt upon in the most reckless man-ner whenever a listener can be found. The fondness for doing such things apparently belongs to the same order of mental obliquity that leads some innocent-minded people to keep dangerous firearms, largely loaded, constantly about the house ; or with as little reason, a ferocious dog in the front yard. Independently of the danger of some one getting and taking such dangerous substances by accident, it should always be remembered that an unnecessary familiarity with such things is of no advantage to any one. While, perhaps, harm-less in some hands, it should never be forgotten that often " the sight of means to do ill deeds, make deeds, ill done."

When it is necessary to get such things, tell the apothecary the use to be made of them, and ask him, as a precautionary measure, to add something to give offensive bulk to the poison without impair-ing its usefulness for the purpose intended.

Further, solutions of Corrosive Sublimate and Oxalic Salt should always be kept in a receptacle the appearance of which, alone, inde-pendent of the name of the substance and the word Poison and other marks, will suggest unpleasant ideas should the contents be used otherwise than legitimately intended.

Several of the commonly kept medicines have individual pecu-liarities not generally known, but it will be so readily seen that the value of the article depends upon them that they will therefore be briefly mentioned in detail.

Alcohol.—This is kept in most houses for various purposes. In a close bottle it will keep for an indefinite length of time.

Aqua Ammoniæ (Water of Ammonia) " Hartshorn."—As the name implies, this is water, saturated, as the chemists say, with a known quantity of Ammonia, a substance, for practical purposes, existing in the form of a gas. The strength, of course, depends upon the quantity of the gas held by the water. As the Ammonia is readily driven from the water by moderate heat, and rapidly leaves it on exposure to the open air, it can be seen that the strength of the

APPENDIX. 115

Aqua Ammoniæ compared with the standard, even when purchased
of the best dealers, may vary a great deal. Owing to the difficulty
of keeping it, a small quantity only should be purchased at one time.
As the ammonia rapidly corrodes the common cork, and finds an outlet for escape from the bottle, the liquid should be kept under a
ground glass stopple.

It is never used internally, but chiefly in combination with other
substances, as a stimulating liniment. If the necessary articles are
to be had, a useful Liniment, when there are no breaks in the skin,
can be easily prepared by mixing equal quantities of Aqua Ammoniæ, Tincture of Opium (Laudanum), Oil of Turpentine, or Glycerine,
or Tincture of Camphor. Any three of these will answer. No
more need be mixed at a time than will last for a few applications.

This Ammonia, united with Carbonic Acid Gas, gives the Carbonate of Ammonia; which, coarsely bruised, and scented with various substances, constitutes the common smelling salts, so much used
by ladies as a nasal stimulant in fainting and hysteria.

Aromatic Spirits of Ammonia.—This is suitably prepared
Aqua Ammoniæ, with other substances. No house should be without this valuable medicine. If the supply can readily be replenished,
no more than an ounce need be bought at once. The value of it as
a medicine depends, of course, upon the useful agent in it, which is
the Ammonia previously spoken of, so care should be taken to keep
up a fresh supply, and take proper care of what you get. If, in an
emergency, it cannot be felt that the article is reliable, the dose may
be increased.

Aromatic Spirits of Ammonia is what physicians call a diffusible
stimulant, in many respects like Brandy, and in some instances
better. The Ammonia in it is an Alkali; so it is also called an
antacid, and given internally to neutralize a supposed excess of acid
in the stomach.

As a stimulant the dose is twenty drops in a teaspoonful of cold
water, every couple of minutes, until certain results sought are obtained.

Tincture of Arnica.—"Arnica." This popular name is as in-
correct as it would be to call Tincture of Opium (Laudanum)
"Opium." It consists of the active principle of the flowers of the
plant, suitably exhausted and kept in solution by Alcohol. It is
useful as a Liniment, by itself or mixed with other things; but perhaps of less value than popularly supposed. It has no more
"healing" properties, when applied to wounds and bruises, than
Laudanum, and is, in fact, not as soothing. As there is no necessity,
unless to soothe pain, for using Laudanum for this purpose, there is
none for pouring on Tincture of Arnica, as many simple-hearted
people do.

Tincture of Arnica, internally, is *poisonous*, like Aconite (see
poisoning by Aconite, p. 93).

Tincture of Camphor.—This is Camphor dissolved in Alcohol
(2 ounces of Camphor to a pint of Alcohol, U. S. P.)

Tincture of Opium.—"Laudanum." This, as the name implies, is the active principle of Opium exhausted by Alcohol. This valuable preparation should always be purchased of a most reliable apothecary. The Tincture of Opium sold in the rural stores often contains scarcely a trace of opium, and may, therefore, be said to be useless. An ounce is enough, as a rule, to keep on hand. If the stopper becomes loosened, from any cause, the Alcohol, of course, evaporates, which might occasion serious inconvenience, by increasing its strength beyond the officinal standard.

The dose of Opium is one grain, and twenty-five drops of the Tincture of Opium (Laudanum) contain this quantity. Hence, twenty-five drops is a dose for an average person in health.

Camphorated Tincture of Opium.—"Paregoric," "Paregoric Elixir." The name suggests its components to a degree; but it contains several other things, none of which interfere with the action of its prominent ingredient—*Opium*.

This should be kept as suggested for the Tincture of Opium (Laudanum).

A tablespoonful (which is half a fluid ounce, two dessertspoonfuls, or four teaspoonfuls, p. 114), contains one grain of Opium. Hence, for an adult a tablespoonful would be a dose; but is rarely used by adults for producing sleep.

A teaspoonful, therefore, contains one-fourth of a grain of Opium, and is equivalent to about six drops of the Tincture of Opium (Laudanum).

It is given to children, however, in preference to Laudanum, for producing sleep. Dose for an infant of one year of age is ten drops; for a child of two years, twenty-five drops.

Camphorated Tincture of Soap, "Soap Liniment," is a well-known and valuable Liniment for ordinary domestic use.

Tincture of Ginger, "Essence of Ginger," is something that everybody buys, but rarely has; because the bottle it comes in is not suitable for keeping it. This tincture should be kept as carefully as the other tinctures named. It should be purchased of the apothecary, but in traveling, when needed, that made by Mr. Frederick Brown, of this city, and sold under the name of "Essence of Jamaica Ginger," can be relied upon.

Oil of Turpentine.—It is said that unless one purchases Oil of Turpentine of the apothecary he is apt to get something more or less adulterated with Benzine. The writer does not know this to be true, but he would suggest that the article should be bought where there can be no doubt of its purity. Owing to the action of light upon this substance it should be kept in a colored bottle, or a bottle pasted over with a piece of thick blue paper. In using it keep away from flame.

Lime Water.—A large bottle of this easily-prepared compound should always be kept prepared for use. It is not only desirable as an antidote to many poisons, as Oxalic Acid, but it is a valuable antacid, when such a thing is required.

To make it, take a piece of unslaked lime (never mind the *size*, because the water will only take up a certain quantity); put it into a perfectly clean bottle and fill the bottle up with *cold* water; keep the bottle corked, and in a cool, dark place, such as a cellar. In a few minutes it is ready for use, and the clear lime-water can be poured off whenever it is needed. When the water is exhausted, fill the bottle again. This may be done three or four times, after which some new lime must be used, as in the beginning.

Compound Syrup of Squill, "Cox's Hive Syrup," improperly called " Syrup of Squill," which is the correct name of another thing, has a well-established use over the country, and is held in esteem by many of the older physicians. Many others, particularly the junior members of the profession, do not like it, from containing a small quantity of Tartar Emetic.

Having in several instances seen a good deal of unnecessary subsequent prostration follow from this ingredient of the Compound Syrup of Squill, the writer feels inclined to recommend the Syrup of Ipecacuanha instead, as an Emetic, leaving the other to be taken when specially ordered by the medical attendant.

Syrup of Ipecacuanha.—This is often known by the popular abbreviation of " Syrup of Ipecac," and when there are children in the house, especially in the winter season, when croup is prevalent, should be kept on hand in quantities of an ounce or so. Children can be made to swallow it easier than they can the pulverized Ipecacuanha or Mustard, when vomiting is required, and from the tendency of small children to dispose of so many things by swallowing them, this syrup is often needed without the time to send far for it. It will keep for some months in a properly secured bottle. The dose, as an emetic, for a child one or two years old, is a teaspoonful or more. This may be repeated every few moments until it acts. In croup and some analogous disorders, the sensibility of the nerves to the stomach appears to be so much impaired, that even this will not act as an Emetic, unless assisted by a Warm Bath.

Mustard.—Pulverized Mustard, or, as it is commonly called, Ground Mustard, should always be kept in every house, and in a place where it can always be found. The kitchen cannot always be depended on for a supply. The time the last is used there is not as often reported, as when some more is needed. The delay caused in sending and getting some, in case of many poisons, often decides the case. It is not easy to get it pure, but if there is any doubt about this, an extra quantity can be given as an emetic, as it will all be rejected, and none is absorbed into the blood. Great care should be observed in keeping the mustard in a tight, wide-mouth bottle, otherwise the delicate active principle will escape from the powder into the air and be lost.

As an E m e t i c, a teaspoonful rubbed down in a teacup of warm water should be given every two or three minutes until vomiting commences, when draughts of warm water should be freely given until there is reason to think the contents of the stomach have been rejected.

Pulverized Ipecacuanha, " Ipecac."—This valuable Emetic, p. 79, should be kept in every house or place where it might be employed. A couple of drachms is enough to get at once, and it should be kept in a bottle with a close-fitting cork. As much as will lie heaped up on an ordinary two cent piece weighs about ten grains.

When it is desired in an attack of Croup to give an E m e t i c, there is nothing better for a child than this substance, as it does not appear to be absorbed into the blood to any extent, or if it does, no harm seems to occur, and it can be given without the fear of giving too much. Even if there should be more given than necessary, like ground mustard, the excess is brought up with the first effort of vomiting. See Syrup of Ipecacuanha, p. 117.

Sulphate of Zinc, " White Vitriol."—Is a prompt Emetic when given in solution in water, in the dose of about twenty grains, as much as will twice lie heaped up on a two cent piece. This should be repeated every few minutes until vomiting follows. Although universally recommended as an Emetic in cases of Poisoning, and when on hand is most valuable, it is always better, instead of waiting for it, to give mustard or common salt. Not that the Sulphate of Zinc is inferior, but because it is so much easier to get ground mustard, and easier still to get common salt, which, in doses of a heaping tablespoonful dissolved in water (four or five teaspoonfuls), is as good as either of the others. It should be given in this quantity every couple of minutes until it acts.

Persons have managed to take an ounce or more of the Sulphate of Zinc for the Sulphate of Magnesia, which its crystals resemble somewhat. If not vomited it would so greatly irritate the stomach and bowels in such a dose, as to entitle it to be called an irritant poison. Half an ounce is enough, therefore, to keep in the house, and it should be kept in a bottle, not a paper.

Seidlitz Powders.—These depend for their value upon their reliability, and this upon the apothecary who sells them. They should be truly made, of active ingredients, freshly compounded and kept perfectly dry in a cool place. Dissolve each powder separately in less than half a tumbler of water; mix together, and drink down while in a state of effervescence.

They should be taken early in the morning, before breakfast, and the water should not be ice-water, for the cold condenses the escaping gas (Carbonic Acid) as it forms, and there is no effervescence.

The components of the Seidlitz Powder, or articles in most respects like them, are combined in various ways as " granular salts," and it is possible, owing to palatability, readiness of carrying, and small bulk, may, in the course of time, as effectually displace the Seidlitz Powder, with many people, as the Seidlitz Powder superseded the less elegant " Epsom Salts," " Rochelle Salt," and other things of the kind.

As these things still maintain a deservedly high reputation, it may be well to say something about them.

Epsom Salt, Sulphate of Magnesia."—The medium dose of this salt, often termed "salts," is an ounce. In bulk this is about two tablespoonfuls.

It should be dissolved in water, no more than is sufficient for the purpose.

In using Epsom Salts, always be sure not to take another substance much resembling it in appearance, and often kept about the house. This is *Oxalic Acid*, a powerful and *rapid poison*. The Oxalic Acid is *sour* to the taste, the Epsom Salt is not.

It also resembles in color, and has been mistaken for the Sulphate of Zinc.

Rochelle Salt, "Tartrate of Potassa and Soda."—The dose of this salt is about half an ounce, or about a tablespoonful.

It should be dissolved in water. It is one of the constituents of the Seidlitz Powder.

Glauber's Salt, "Sulphate of Soda."—This old-fashioned, disagreeable salt is gradually disappearing from use, having been superseded by the Sulphate of Magnesia (Epsom Salt) which is more agreeable to take.

The dose is half an ounce, dissolved in water, and taken like all other medicines of the class, of a saline character, upon getting up, before breakfast. A little lemon-juice, or a pinch of cream of tartar, is said to make it more acceptable to the taste.

Pills.—The persistent use of cathartics, whether in shape of pills, salt, or liquid, is sure to bring on trouble which nothing else for the rest of the life may correct. Be careful of what you eat, how you eat, and when you eat, and in a few months, never mind how much medicine of the kind is taken now, you will find, unless it is an exceptional case, a marked improvement in health.

If medicines, particularly pills, must be taken, do not use the wretched "vegetable" varieties, when there are so many better, to be had of any Apothecary, freshly made, of the best material, in proper quantities, and, as some people will say, at a much less price. The physician in the country, too, makes the same thing, in the same way, and under the same name. There are several of them, but the two to be remembered are the

Compound Cathartic Pill

Compound Rhubarb Pill.—Three or four of either are a safe, gentle cathartic, acting by morning, when taken at late bedtime. Two will often answer.

These two kinds of pills being made according to the precise directions of the U. S. Pharmacopœia, can be had of any Apothecary, in any quantity; and as they are made of the best material, they may be relied upon for the purpose intended. Beside these advantages, they are so constantly ordered by Physicians in prescriptions, that they are usually freshly made,

INDEX.

Portions of bodies cleanly cut off, 53.
Potash, internally, 90.
Powders, how to preserve, 112.
Precautions to be observed in hot weather, 23.
" Proud flesh," granulations, 57.
Pulsatilla, 103.
Pulverized Ipecacuanha, 117 ; as an emetic, 79.
Punctured wounds, 53; how to treat, 54; dangers to be appre-
hended from, 55.
Pus; formation of, 57 ; discharge of, 54 ; effects of excessive forma-
tion on strength, 32.

" Red-precipitate," as a poison, 89.
Rochelle salt, 118.
Rubbing with broken ice, 21.

Sanguinaria, 103.
Scalp-wounds, 58.
Salt, common, as an emetic, 78.
Seidlitz powders, 118.
Shaving the face, cuts, what to do, 49.
Shock, or collapse, 25 ; symptoms of, 24; in burns and scalds, 29 ;
in contusions, 34 ; what to do, 25–28 ; confounded with faint-
ing, 25 ; death from, 25 ; in age and infancy compared, 26 ;
from bathing in, or drinking cold water, 28.
Shell-fish, poisonous, 82.
Silver, salts of, as poisons, 91.
Signs of real death, 106–113.
Soap and water, object in using for wounds, 52.
Spigelia, 103.
Squill, compound syrup of, 117.
Soda, as a poison, 91.
Splinters ; in hand, 55 ; under the nail, 55 ; how to remove, 55.
Stimulants; ammonia and brandy (note), 22 ; effects of too much, 47.
Stings of insects, 59.
Stomach pump, 76.
Stretcher, how used, 7.
Stramonium, 104.
Strychnia, 104.
Sunstroke, 18–24 ; general observations on, 19 ; symptoms, 19 ;
treatment, 20 ; prevention, 22.
Suffocation, 8–18.
Sulphuric acid, "Oil of Vitriol," burns by, 33 ; poisoning by, 83.
Sulphate of Magnesia, 119 ; of Soda, 119 ; of Zinc, 118 ; of Zinc, as
an emetic, 79 ; of Zinc, poisoning by, 91.
Sulphuretted Hydrogen, 18.
Suppuration, effects of on strength, 32.
Sutures, 50.
Syrups, how to preserve, 112.
Syrup of Ipecacuanha, 117 ; of Squill, 117.

www.ingramcontent.com/pod-product-compliance
Lightning Source LLC
Chambersburg PA
CBHW021937190326
41519CB00009B/1049